PHYTOREMEDIATION AND STRESS: EVALUATION OF HEAVY METAL-INDUCED STRESS IN PLANTS

BOTANICAL RESEARCH AND PRACTICES

Additional books in this series can be found on Nova's website under the Series tab.

Additional E-books in this series can be found on Nova's website under the E-book tab.

PHYTOREMEDIATION AND STRESS: EVALUATION OF HEAVY METAL-INDUCED STRESS IN PLANTS

JANA KADUKOVA
AND
JANA KAVULICOVA

Nova Science Publishers, Inc.
New York

For permission to use material from this book please contact us:
Telephone 631-231-7269; Fax 631-231-8175
Web Site: http://www.novapublishers.com

NOTICE TO THE READER

Additional color graphics may be available in the e-book version of this book.

LIBRARY OF CONGRESS CATALOGING-IN-PUBLICATION DATA

Kadukova, Jana.
  Phytoremediation and stress : evaluation of heavy metal-induced stress in plants / authors: Jana Kadukova and Jana Kavulicova.
     p. cm.
  Includes index.
  ISBN 978-1-61761-319-7 (softcover)
  1. Plants--Effect of heavy metals on. 2. Plants--Effect of stress on. 3. Phytoremediation. I. Kavulicova, Jana. II. Title. III. Title: Evaluation of heavy metal-induced stress in plants.
  QK753.H4K33 2010
  571.9'2--dc22
                                                                              2010036419

Published by Nova Science Publishers, Inc. † *New York*

CONTENTS

.

PREFACE

Phytoremediation advantages are widely known nowadays. It is a method applicable for treatment of large areas with low concentration of pollutants or areas where only a finishing step of cleaning is required. Very often these kinds of places represent great problems because there is no possibility to take all the soil to the landfills and often they are part of agricultural fields. There are many studies dealing with application of variety of plants for treatment of soils contaminated by heavy metals or organics.

The book is focused on heavy metal contamination cleaning up by phytoremediation with the aim to describe some of the possibilities how to assess the stress of plants. Phytoremediation, as a technology, needs credible markers enabling us to evaluate the stress induced by conditions where plants are growing. There are several parameters which can be used in the plant stress assessment such as reduction of biomass production, plant growth inhibition, changes in photosynthesis, germination inhibition, and production of antioxidant enzymes. In regard to differences in plant responses to heavy metal-induced stress especially in the state of antioxidative systems it is necessary to prepare a standard protocol for the evaluation of the impact metals have on plant health but generally for stress evaluation it is important to assess several "stress indicators" together. Combination of parameters expressing the influence on photosynthesis and cell oxidative status would be desirable. Knowledge of these factors can bring us closer to understanding of molecular mechanisms of heavy metal accumulation by plants and it indirectly can help further application of phytoremediation as well as has numerous additional biotechnological implications.

INTRODUCTION

Accumulation of different compounds from soil, water or even from the atmosphere is a typical feature of plants as the organisms dependent on mineral nutrition. However, in the case where high concentrations of metals are present in the soil their accumulation is not always considered good or necessary, especially in medicinal plants or agricultural crops. But even this "bad feature" can be used for a human benefit. Nowadays, phytoremediation, biotechnology using plants for cleaning up the environment, has become widely applied. Its development can contribute to re-exploitation of soils that have laid waste due to their contamination. Similarly, plants able to tolerate elevated metal concentrations without their translocation and accumulation in upper parts can grow on contaminated soils not threatening food chain. These soils are then not only utilized but also stabilized and thus preventing environment from circulation of soil contaminants. Further development of phytoremediation requires an integrated multidisciplinary research effort that combines plant biology, soil biochemistry, soil microbiology as well as agricultural and environmental engineering.

Using plants in phytoremediation is specific because it takes account of plant application in stressful conditions in advance. So it is necessary to know not only the ability of plants to accumulate metals but also their ability to cope with stress. Dealing with phytoremediation finally requires the way of expressing the extent of the heavy metal-induced stress in numbers to be able understand better what is happening at the level of plant and to be able to evaluate suitability of the particular plant for special conditions, as well as to compare results published in scientific literature.

The concept of heavy metal-induced stress is rather wide. As at the molecular level it is a complex process, it is difficult to reduce it to individual measurable components. Nevertheless, it is possible to select parameters expressing the extent of stress for plant. In the book we have concentrated on some of those which characterize the metal influence on photosynthetic apparatus of the plant cells as well as the degree of oxidative stress and activation of mechanisms responsible for metal accumulation.

The aim of this book is to describe commonly measured parameters in plants exposed to metals and compare their expressive value facilitating stress evaluation in plants growing in the presence of excessive amounts of heavy metals. We have tried to summarize available information about the influence of metals on plant organism from the view of the way of metal sequestration, their accumulation and translocation in plant bodies as well as from the view of the negative effects of metals and stress elicitation.

Chapter 1

PHYTOREMEDIATION

According to Mench et al. (2010) it is estimated that there may be up to 3 million potentially contaminated sites in EU including approximately 250 000 sites which need urgent attention. Contamination by metals accounts for more than 37% of cases, followed by mineral oil (33.7%) and polycyclic aromatic hydrocarbons – PAH (13.3%) and others (Vamerali et al., 2010). Analyzing the situation in Europe and USA Lewandowski et al. (2006) found that approximately 100 000 ha of land are contaminated by heavy metals.

But because of the potential toxicity and high persistence of heavy metals, the clean up of contaminated soils is one of the most difficult tasks for environmental engineers (Wu et al., 2004). Very often contaminated soil is treated as waste and it is disposed. There are several physico-chemical methods used for treatment of contaminated soils, such as soil washing (extraction of metal by organic or inorganic acid), soil removal, capping (Nedelkoska, Doran, 2000). These methods are rather expensive costing from 50 to 560 €.t^{-1} of soil (Witters et al., 2009).

In-situ remediation methods are desirable for soil since it is a valuable resource that can be renewed only slowly (Warren et al., 2003). One of perspective methods in heavy metal contaminated soil treatment includes phytoremediation. Phytoremediation has been defined as the use of green plants and their associated rhizospheric microorganisms, soil amendments, and agronomic techniques to remove, degrade, or detoxify harmful environmental pollutants (Ouyang, 2002, Schwitzguébel, 2002). It offers efficient tools and environmentally friendly solutions for soil clean-up contributing to soil restoration and sustainable land use management. It does not interrupt land use during and after remediation or reduce soil fertility like conventional methods.

The cost of phytoremediation can be compared with regular farming activities (Witters at el., 2009).

First mention of plant accumulation of high amounts of metals was described in 1885 by Baumann for *Thlaspi caerulescens* and *Viola calaminaria* (Lasat, 2000) and slowly increased knowledge of this phenomenon. But it took almost one hundred years until it was used as an environmental counterbalance to industrial processes. The idea of using plants that hyperaccumulate metals to selectively remove and recycle excessive soil metals was introduced in 1983, gained public exposure in 1990, and has increasingly been examined as a potential practical and more cost-effective technology than the soil replacement, solidification and washing strategies presently used (Chaney et al., 1997). The term phytoremediation was used for the first time in 1991 and means plant-based action (*phyto* – plant, *remediation* – to recover) (Raskin et al., 1994).

Plants ideal for phytoremediation should fulfill four main requirements (Schnoor, 1997):

1. they must be fast growing and produce high biomass,
2. have deep roots,
3. have easily harvestable aboveground portion,
4. accumulate large amounts of metals (~ 1000 mg/kg) in aboveground biomass.

But in fact there is not known plant which would fulfill all these requirements. Most of the heavy metal accumulating plants have only shallow roots and produce small biomass (Yang et al., 2005). In addition they often accumulate only one specific element (Doumett et al., 2008). Due to these reasons except of using metal-hyperaccumulating plants there are also other strategies in plant use for soil treatment. One of them is to use the fast growing plants (producing high biomass) accumulating low or average amount of metals or to use high biomass crops together with chemical and microbial treatment (Yang et al., 2005, Lievens et al., 2008). The use of selected trees or shrubs that are fast growing, have deep root system and are able to grow on nutrient-poor soil is also a perspective alternative (Pulford, Watson, 2003).

Phytoremediation represents many advantages compared to other remediation techniques (Raskin et al., 1994, Schwitzguébel, 2002, Dercová et al., 2005):

*   it can be performed with minimal environmental disturbance;

- it is applicable to a broad range of contaminants, including many metals with limited alternative options and radionuclides;
- possibly less secondary air and water wastes are generated than by traditional methods;
- organic pollutants may be degraded to CO_2 and H_2O, removing environmental toxicity;
- it is cost-effective for large volumes of water having low concentrations of contaminants;
- topsoil is left in usable conditions and may be reclaimed for agricultural use;
- soil can be left at the site after contaminants are removed, rather than having to be disposed or isolated;
- it is cost-effective for large areas having low to moderately contaminated surface soils;
- plant uptake of contaminated groundwater can prevent off-site migration;
- does not need to use heavy vehicles and devices which damage soil.

However, there are also several drawbacks and limitations of phytoremediation also:

- there is a long time often required for remediation;
- the treatment is generally limited to soils contaminated up to one meter from the surface;
- climatic or hydrologic conditions may restrict the rate of growth of plants that can be utilized;
- the ground surface at the site may have to be modified to prevent flooding or erosion;
- contaminants may still enter the food chain through animals/insects that eat plant material containing contaminants;
- soil amendments may be required.

Phytoremediation is best applied at sites with shallow contamination of organic, nutrient, or metal pollutants. It is well-suited for use at very large field sites where other methods of remediation are not cost-effective or practicable (Schnoor, 1997). Plants can be used to treat most classes of contaminants - toxic metals, radionuclides and recalcitrant organic pollutants, like chlorinated pesticides, organophosphate insecticides, petroleum hydrocarbons (BTEX), polynuclear aromatic hydrocarbons (PAHs), sulfonated aromatics, phenolics, nitroaromatics and explosives, polychlorinated biphenyls (PCBs), and chlorinated solvents (TCE, PCE). This method is often complementary to traditional bioremediation techniques, based on the use of microorganisms only (Alkorta, Garbisu, 2001, Ouyang, 2002, Schwitzguébel, 2002, Abhilash et al., 2009). Remediation of inorganic contaminants differs from the case with organic compounds. Because organic compounds can be mineralized, but the remediation of inorganic contamination must either physically remove the contaminant from the system or convert it into a biologically inert form (Cunningham, Ow, 1996). But even in the case that there is no other possibility as to landfill harvested plants (e.g. metals accumulated in plant body have no use or there is no way how to economically extract metal from the plant, etc.) phytoremediation brings advantages because of significant reduction of contaminated material for landfilling. Besides soil treatment plants have been successfully used for treatment of wastewater (municipal and industrial wastewater) (Majer Newman et al., 2000, Dunne et al., 2005, Chavan et al., 2007, Zurita et al., 2009, Khan et al., 2009) and even some information is known for treatment of the atmosphere (Liua et al., 2007).

1.1. PHYTOREMEDIATION TECHNIQUES

With the increasing amount of information about phytoremediation and development of new applications several phytoremediation techniques can be distinguished (Schwitzguébel, 2002, Pulford, Watson, 2003, Yang et al., 2005, Mackova et al., 2006, Gerhardt et al., 2009). The most common are listed in Table 1.

Phytoremediation techniques suitable for heavy metal contaminated soils or water cleaning up basically include phytostabilization, phytoextraction, rhizofiltration and phytovolatilization. Rhizospheric microorganisms play an important role in metal phytoremediation (Ike et al., 2007).

Phytostabilization is usually applied on heavily contaminated soils so that removal of metals using plants would take an unrealistic amount of time. In such a case it is the best to choose fast growing plants which can grow in metal contaminated and nutrient deficient soil and which can immobilize heavy metals through absorption and accumulation by roots or precipitation within the rhizosphere not translocating them into shoots (Wong, 2003, Pietramellara et al., 2009). It was found that significant fraction of metals can be stored at the root level, especially in polyannual species contributing to long-term stabilization of pollutants (Vamerali et al., 2010). At present, phytostabilization is often used to revegetate mine tailings to minimize wind and water erosion of tailings in a cost-effective way. Phytostabilization also improves the chemical and biological characteristics of the contaminated soils (Alvarenga et al., 2008, Chen et al., 2008, Grandlic et al., 2009).

Table 1. Phytoremediation techniques

Technique	Characteristics
phytoextraction (phytoaccumulation)	the use of pollutant-accumulating plants to remove metals or organics from soil by concentrating them in harvestable plant parts
phytotransformation	the partial or total degradation of complex organic molecules or their incorporation into plant tissues
rhizoremediation (plant-assisted bioremediation, phytostimulation)	the release of plant exudates/enzymes into the root zone stimulates the microbial and fungal degradation of organic pollutants
rhizofiltration	the use of plant roots to absorb or adsorb pollutants, mainly metals, but also organic pollutants, from water and aqueous waste streams, concentrate and precipitate them
phytostabilization	the use of plants to reduce the mobility and bioavailability of pollutants in the environment, preventing their migration to groundwater or their entry into the food chain
phytovolatilization	the use of plants to volatilize pollutants or metabolites
removal of aerial contaminants	uptake of various volatile organics by leaves
dendroremediation	the use of trees to evaporate water and to extract pollutants from the soil
hydraulic control	the control of the water table and the soil field capacity by plant canopies.

Phytoextraction is a method used for extracting metals from the soils by concentrating them in harvestable plant parts. It can be divided into two methods – induced phytoextraction and continuous phytoextraction (Salt et al., 1998). Induced phytoextraction uses plants producing big amount of biomass which metal accumulation is enhanced by addition of chemicals, such as EDTA, NTA, EDDS etc. (Kos, Leštan, 2003, Quartacci et al., 2007, Saifullah et al., 2009). Continuous phytoextraction uses plants with natural abilities to accumulate high levels of metals – hyperaccumulators (McGrath et al., 2002). Hyperaccumulating plants have an extraordinary ability to accumulate heavy metals, translocate and concentrate them in roots and above ground shoots or leaves (Schnoor, 1997, Clemens et al., 2002). Metal translocation from the roots to the shoots for the purpose of harvesting is one of the key goals of phytoextraction research (Jarvis, Leung, 2001). An important issue in phytoextraction is whether the metals can be economically recovered from the plant tissue or whether disposal of the waste is required. Several methods have been studied for contaminated plants disposal such as incineration, liquid extraction, ashing or direct disposal at a hazardous waste site. Their application depends on the cost of the process and availability of the appropriate technology (Sas-Nowosielska et al., 2004)

Interesting possibilities are brought by the use of plants with special characteristics to make the process more efficient. For example, Mediterranean halophytic shrub or tree *Tamarix smyrnensis* uses salt glands to excrete the excess of salt from soil. In the case that metals are present in soil these are excreted through salt glands in the form of non-toxic crystals. This method is called **phytoexcretion** (Kadukova, Kalogerakis, 2007, Manousaki et al., 2008).

Rhizofiltration is primarily for treatment of extracted groundwater, surface water, and wastewater. It is defined as the use of plants, both terrestrial and aquatic, to absorb, concentrate, and precipitate contaminants from polluted aqueous sources in their roots (Schwitzguébel, 2002, Eapen et al., 2003, Peng et al., 2008). The advantage of rhizofiltration is that contaminants do not have to be translocated to the shoots, thus, species other than hyperaccumulators may be used. This method is suitable for treatment of water with low metal contamination (Schwitzguébel, 2002, Rousseau et al., 2004, Choo et al., 2006) and for treatment of water contaminated with radionuclides (Soudek et al., 2007, Vera Tomé et al., 2008).

Phytovolatilization involves the use of plants to take up contaminants from the soil, transforming them into volatile forms and transpiring them into the atmosphere (Schnoor, 1997). Genetically modified *Arabidopsis thaliana* and *Lyriodendron tulipifera* can grow in soil with higher mercury concentration and transfer it from Hg^{2+} to Hg^0 (Špirochová et al., 2001). Using this method is necessary to control release of formed volatile compounds to the atmosphere.

Selection of plants for phytoremediation of metals depends on the particular application - for phytostabilization, rhizofiltration, phytovolatilization or phytoextraction. Plants that accumulate toxic metals can be grown and harvested economically, leaving the soil or water with a greatly reduced level of toxic metal contamination. Dried, ashed or composted plant residues, highly enriched in heavy metals may be isolated as hazardous waste or recycled as bio-metal ore (Raskin et al., 1994, Koppolu et al., 2003).

The use of plant in metal recovery is an emerging technique. This phytotechnology is called **phytomining** and it is the special case of phytoextraction. There are several results published about the high accumulation ability of some plants to accumulate gold and silver. These results open the future possibility to use the plants not only to clean the environment but also to recover specific metals from soil (Gardea-Torresdey et al., 2005).

1.2. PRACTICAL APPLICATION OF PHYTOREMEDIATION

Phytoremediation has been a really intensively studied technology for over last 15 years. Different information has been gathered during this time related to metal accumulation by plants from laboratory or greenhouse research. Although this kind of research helped a lot in the understanding of heavy metal accumulation by plants, it is difficult to decide if all the results are applicable for fields and large contaminated areas. Factors influencing soil conditions, metals speciation and their bioavailability are different in soil in pots and in the fields (Zabłudovska et al., 2009, Mench et al., 2010). For example, uptake of metals by plant is typically 1 – 2 orders of magnitude lower when grown in soil compared with the uptake by the same plant grown in culture solution (Kayser et al., 2000). To extend the phytoremediation more field studies will be necessary to carry out for understanding the factors responsible for the process in natural condition. If these factors were optimized, the potential for phytoremediation use could be increased (McGrath et al., 2006). In every case, widening application of phytoremediation techniques is obvious at present. Glass (2000) reported that only in the USA the assets in phytoremediation had increased since 1997 from 1 - 2 $ to 15 – 25 $ in 2000. The public acceptation of phytoremediation helping its further application is the great advantage.

For the practical and economical phytoremediation application not only plant capacity to accumulate metal or rate of biomass production but also other processes and parameters such as necessary agronomic technique, ground water capture zone, contaminated plant material treatment etc. are important.

At present *phytostabilization* and *phytoextraction* are very often used phytoremediation methods to remove heavy metals from contaminated soils. With the depleting of metal resources *phytomining* started to gain attention as well. Simplified decision-making process is shown in the Fig. 1.

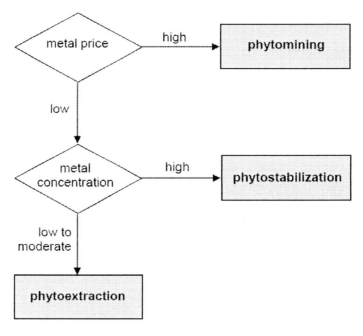

Figure 1. Decision-making process in using phytoremediation techniques for metal contaminated soils treatment.

1.2.1. Phytostabilization

Phytostabilization, as it was describe earlier, is the method using plants to reduce the mobility and bioavailability of pollutants in the environment. The aim of the method is to stabilize heavy metals in solid matrices through mechanical and bio(chemical) processes preventing them to migrate and circulate within the environment and food chain (Vamerali et al., 2010). This method. is used when other phytoremediation techniques, mainly phytoextraction, would last for a very long time (e.g. on heavily contaminated soils) or no alternative treatment is available. So it is not exactly the technology for soil cleaning-up but more a management strategy for contaminants stabilizing consequently reducing the risks presented by contaminated soil (Vangronsveld et al., 2009).

Phytostabilization is in practice often applied as aided phytostabilization. Firstly single or combined amendments are incorporated into soil to decrease the labile contaminant pool and phytotoxicity finally allowing revegetation of contaminated sites (Pérez-de-Mora et al., 2007, Mench et al., 2010). In contrast to phytoextraction plants used for phytostabilization should not accumulate high concentrations of heavy metal into their aboveground parts because this would facilitate their entry into food chain (Domínguez et al., 2008). Usually it is only a temporary solution until better soil remediation technology for the particular site will be developed. Plants covering the contaminated area decrease the wind and water erosion as well as metal leaching by returning rainfall to the atmosphere via evapotranspiration and so protecting groundwater (Tordoff et al., 2000, Ruttens et al., 2006a). Both the improvement of aesthetic value of formerly bare areas and the improvement of chemical and biological characteristics of contaminated soil by increasing the amount of organic matter, nutrient levels, cation exchange capacity and biological activity are also important (Arienzo et al., 2004, Ruttens et al., 2006b).

The main requirement for plant selection is not the metal accumulating ability as it is in the case of phytoextraction but the ability to tolerate high concentration of heavy metals in soil as well as the ability to develop extended and abundant root systems (Mendez, Maier, 2008). Plants in cooperation with soil bacteria and mycorhiza living in the rhizosphere are responsible for several processes during phytostabilization (Rizzi et al., 2004, Vangronsveld et al., 2009):

- metal accumulation and precipitation in roots or root zone
- metal adsorption on root surfaces
- changes in metal speciation due to changes of soil characteristics (e.g. pH, Eh, …).

The amendments added into soil during aided phytostabilization bind metals (through sorption or precipitation) so they restrain the metal ion migration from soil into other parts of the environment as well as decrease metal bioavailability and consequently their toxicity to plants (Warren, Alloway, 2003, Ruttens et al., 2006b). They also increase soil fertility to restore function to the ecosystem (Brown et al., 2005). The most commonly lime, phosphates, biosolids, natural and synthetic zeolites, sludges and/or compost are used as amendments (Gray et al., 2006). However, some of them can have an undesirable side effect, such as destroying of soil structure or immobilizing of essential nutrients (Vangronsveld et al., 2009).

Phytostabilization Field Case Studies

Guadiamar River Valley, Spain

After the toxic sludge spill (cca. 4 million m^3) from mine tailing dam at Aznalcóllar approximately 55 km^2 of Guardian River basin was contaminated with high quantities of As, Cd, Cu, Pb, Tl, Zn, Sb and Bi. To fix metals in soil and reduce their negative impact on the environment and prohibit the further expansion of the contamination phytostabilzation was used. Organic matter and Ca-rich amendments were added with the aim of immobilizing trace metals and improving soil fertility. Shrubs and trees typical for the area were used for the revegetation: *Populus alba, Fraxinus angustifolia, Quercus ilex* subsp. *hallota, Olea europaea* var. *sylvestris, Tamarix africana, Salix atrocinerea, Phillyrea angustifolia, Pistacia, lentiscus, Rosmarinus officinalis* and *Retama sphaerocarpa*. Despite the high metal concentration in soils, there was limited transfer of these elements to the aboveground parts of woody plants. Only exception represented *Populus alba* but even in this plant ecological risk of the release of metals from leaves was low (Domínguez et al., 2008).

In the area also the phytoextraction study was carried out using *Brassica juncea* but because of low phytoextraction capacity unrealistic long time would be necessary to reduce metal concentration. Phytostabilization studies are more realistic in the area (Clemente et al., 2005).

Sanlúcar la Mayor, Seville, Spain

The area was affected by the same accident as the previous one (the toxic sludge spill from Aznalcóllar pyrite mine). *Lupinus albus* was used for the phytostabilization of acid polluted soil containing elevated levels of Cd and As. During the study it was found that *Lupinus albus* in cooperation with rhizospheric organisms decreased the soluble As and Cd fractions in soil mainly because of the increase of As and Cd contents in plants and because of the increase of soil pH. Most of the metals were stored in roots and root nodules strongly bound to cell walls. Therefore, the culture of white lupine plants on soils have not only benefits on the recovering of pH status but also reduced the soluble levels of As and Cd (Vázquez et al., 2006).

Torviscosa, Udine, Italy

The site was polluted with pyrite cinders produced during sulfur extraction. At the site a 0.7 m layer of As (886 $mg.kg^{-1}$), Co (100 $mg.kg^{-1}$), Cu (1735 $mg.kg^{-1}$), Pb (493 $mg.kg^{-1}$) and Zn (2404 $mg.kg^{-1}$) contaminated wastes had been covered with an unpolluted 0.15 m layer of gravelly soil. At the beginning of the experiment soil was fertilized with N. P, K fertilizers. Two-year-old bare rooted cuttings of *Populus alba, P. nigra,* and *Salix alba* and one-year-old cuttings of *Populus tremula* were used in the study. Uptake of the metals and As to aboveground plant parts was determined as marginal so there was a little risk of food chain contamination via herbivores. Besides, vegetation cover was formed reducing wind-blow of soil and playing an important landscape and ecological role (Vamerali et al., 2009).

1.2.2. Phytoextraction

Phytoextraction is the use of heavy metal-accumulating plants to remove metals from soil by concentrating them in harvestable plant parts. Hyperaccumulators are very often used in this technique (Murakami, Ae, 2009). In terms of phytoextraction metal accumulation is usually expressed by the metal biological absorption coefficient – BAC (the ratio of metal concentration in harvestable plant part to soil metal concentration). Besides BAC, both the bioconcentration factor – BCF (the ratio of metal concentration in roots to soil metal concentration) and the translocation factor – TF (the ratio of metal concentration in shoot to metal concentration in roots) are used in phytoextraction studies (Vamerali et al., 2010). The plants with high BAC (higher than 1) are the most suitable for phytoextraction. The plants with high BCF and low TF are more suitable for phytostabilization (Yoon et al., 2006).

Phytoextraction as a technology represents repeated cropping of plants on heavy-metal contaminated soils until the soils' metal concentrations have reached acceptable level (Robinson et al., 2003). Phytoextraction involves three basic steps:

1. cultivating of metal accumulating plants on contaminated soil,
2. harvesting of the plants after specific time (weeks or months),
3. post-harvesting treatment of plant biomass (combustion, microbial degradation, etc.).

Climatic conditions of the site influence the efficiency of the method. The soil constituents (and their proportions in soil) may be critical information for selection of soils with high potential for remediation through phytoextraction (Hammer et al., 2006). In general, three strategies are used (McGrath, et al., 2006, Maxted et al., 2007, Chaney et al., 2007):

1. cultivation of arable crops with or without application of heavy metal mobilizing agents,
2. cultivation of fast growing plants, shrubs and trees accumulating low or average amount of metals but producing high amount of biomass (heavy metal mobilizing agents can be used also in this case) – this process is called induced phytoextraction,
3. cultivation of metal-hyperaccumulating plants.

When shrubs or trees are used for phytoextraction, metal can return back to soil via leaf-fall resulting in concentrating even more metal in the upper soil profile (Robinson, et al., 2003). Critical parameters limiting the use of phytoextraction include plant biomass yield and portion of metal accumulated in harvestable plant parts (Puschenreiter et al., 2001). Phytoextraction appears to be a balance between the phytotoxic effect of metals and their uptake by plants (Schwartz et al., 2001).

Phytoextraction as a technology is more realistic for treatment of soil containing only moderately elevated metal concentration found mostly in soil used in agriculture where people are endangered due to long-term and repeating crop cultivating. For example Blaylock (2000) has calculated that Pb extraction by *Brassica juncea* is feasible only for sites of which metal concentration did not exceed 1500 mg.kg^{-1}. The sources of such pollution are usually long-term utilization of synthetic fertilizers, utilization of sewage sludges as soil amendments or pollution of soil because of atmospheric pollution from nearby industrial factories (Maxted et al., 2007, Lebeau et al., 2008). In the case of a very high metal contamination phytoextraction would take too long (Clemente et al., 2005, Meers et al., 2005). The reasonable period for soil phytoremediation is considered to be less than 5 years (Khan et al., 2000). It is only a suitable technique when contamination is located in the upper layer of the soil, within the reach of the roots, which on average is less than 50 cm (Witters et al., 2009). But in every case even low and moderately polluted soils can make problems to farmers because they cannot use such soil for farming of valuable crops and they have to set the land aside or switch from the high value vegetable production to the production of cereals that generates a lower gross margin (Lebeau et al., 2008). In these cases phytoextraction is usually the only suitable technology.

To avoid phytoextraction limitations Meers et al. (2010) suggested using high biomass crops which, after phytoextraction, would be further valorized. This would generate the alternative income for farmers and soil decontamination would come at the second level. This approach could help especially in regions where cultivation of food or feed crops is impossible because of the presence of high concentrations of plant available metals.

Phytoextraction Field Case Studies

Nottingham, UK

The study was carried out at sewage sludge processing facility near Nottingham, UK. The whole contaminated area was a size approximately 630 ha of agricultural soil used for production of livestock feed. Smaller plots were chosen for the Cd and Zn phytoextraction study. The total Cd and Zn concentration in soil was 34.5 and 2160 mg.kg^{-1}, respectively. A hyperaccumulating species *Thlaspi caerulescens* was used for Cd and Zn extraction. Biomass yield of was from 0.5 to 4 t.ha^{-1}. Cd and Zn concentrations in plants were in average 124 and 980 mg.kg^{-1}, respectively. Repeating the experiment, some differences were found probably due to intrinsic variation in the physiological and genetic characteristics of seeds collected from a wild population. Authors of the study suggested that repeated and multiply croppings within a single season would be required for successful phytoextraction (Maxted et al., 2007).

St. Petersburg Region, Russia

In the field study U and Th was extracted using two native species *Elytrigia repens* and *Plantago major* and two cultivated plants *Triticum aestivum* and *Secale cereale*. Uranium concentration in studied soil varied from 0.5 to 5.5 mg.kg^{-1}, thorium concentration was in a range 5.6 – 10.7 mg.kg^{-1}. Although plants were able to accumulate significant amount of metals they were stored mostly in roots not translocated into shoots. These plants would be probably more suitable for phytostabilization of radionuclides (Shtangeeva, 2010).

La Bouzule, Lorraine, France

Thlaspi caerulescens plants were used to clean up agricultural soil amended with heavy metal rich urban sludge containing 24 mg.kg^{-1} of Cd and 2347 mg.kg^{-1} of Zn. Two crops of *T. caerulescens* extracted about 9% of the total cadmium added into soil as contaminated sewage sludge and simultaneously 7% of the total Zn. A higher uptake was recorded with the larger plants (Schwartz et al., 2003).

Woburn Market Garden, Bedfordshire, UK

The place was contaminated because of long-term use of varying amount of sewage sludge. Concentrations of Zn and Cd exceeded legal limits accepted for agricultural soil in UK or approached the maximum concentrations allowed. Seedlings of *Thlaspi caerulescens* and *Arabidopsis halleri* were planted in 4 m^2 area. The effect of EDTA, NTA and citric acid was examined during experiments. It was found that *A. halleri* produced lower biomass that *T. caerulescens* resulting in lower metal extraction. *A. halleri* could only extract up to 0.004 kg.ha^{-1} Cd and 0.63 kg.ha^{-1} Zn while *T. caerulescens* extracted up to 0.7 kg.ha^{-1} and 4.1 kg.ha^{-1} of Cd and Zn, respectively. Hyperaccumulation capacity of *T. caerulescens* for Cd is so high that even with low biomass production it is more efficient Cd accumulator than high biomass crops with low Cd accumulation capacity. The results in the study indicated that none of the metal mobilizing chemicals used were able to enhance Cd and Zn hyperaccumulation by *T. caerulescens* (McGrath et al., 2006).

Guangzhou, China

Arable agricultural land of approximately 600 ha which received the sediment from the Pearl River was chosen for the study (only 1500 m^2 for the real field study). A fruit tree *Averrhoa carambola*, which can reach the height of about 10 m, was used for Cd extraction. Cd concentration in soil before treatment was 1.6 mg.kg^{-1}. This plant produced during 170 days 23 t.ha^{-1} of total biomass. After that time 5.3% of Cd from soil was removed. To reduce Cd concentration to the limit 1 mg.kg^{-1} would take approximately 9 years. But even the Cd concentration reduction after the first year of phytoextraction study was high enough to reduce Cd level in the leaves of vegetable *Brassica parachinensis* under permissible concentration for vegetable in comparison with *B. parachinensis* grown in untreated soil (Li et al., 2009).

Bazoches and Toulouse, France

Field experiments were carried out in the soil surrounding Pb recycling plants near to Bazoches (50 m^2 field) and Toulouse (10 m^2 field). For the phytoextraction study the cultivars of *Pelargonium* were used. But even with the biomass yield 45.3 tons.ha^{-1} and Pb-hyperaccumuating capacity of one of the selected *Pelargonium* cultivars approximately 181 years would be necessary to clean up the soil. According to results *Pelargonium* plants could be a suitable alternative for less polluted soils (Arshad et al., 2008).

Lommel, Belgium

The study took place on a former maize field with following metal concentrations: 4.1 – 7.4 mg.kg^{-1} of Cd, 160 – 222 mg.kg^{-1} of Pb and 210 – 418 mg.kg^{-1} of Zn. Two species of energy crops (*Salix* spp. and *Populus* spp.), *Zea mays* for biogas production and *Brassica* sp. for pure plant oil production were used in the study. The best results with a short-rotation coppice of willow (*Salix spp.*) were reached (Witters et al., 2009).

1.2.3. Phytomining

Phytomining is the technology which uses plants to recover metals from metalliferous soils or low–grade ore bodies (Sheoran et al., 2009, Bali et al., 2010). The difference between phytomining and phytoremediation represents the aim of the technology. While the aim of phytoremediation, namely phytoextraction, is to clean up the contaminated environment, the aim of phytomining is to gain metals but the practical application is similar. Anderson et al. (2005) suggested using preferably the term "phyto-reclamation" to avoid misunderstandings as phytomining is not an alternative to a conventional mining. The growing interest in phytomining results from increased societal pressure connected with the environmental performance of conventional mining processes, the inability of current technologies to cost effectively recover metals from ores with low metal contents as well as high metal prices (Harris et al., 2009).

Phytomining operation would entail planting a hyperaccumulator crop over metalliferous soil and then harvesting. Dried plant material is then burned into an ash with or without energy recovery. A commercial "bio-ore" is produced by the process which can be further processes by conventional metal refining methods such as roasting, sintering, smelting or chemical leaching from the biomass (Robinson et al., 1999, Anderson et al., 1999, Barbaroux et al., 2009). It offers the possibility of exploiting ore bodies or mineralized soils which would be uneconomic by other methods (Boominathan et al., 2004). Advantages of this method include also higher metal content of "bio-ore" and lower sulfur content (even if it is produced on the sulfidic ore containing substrate) so smelting a "bio-ore" does not contribute significantly to acid rain (Brooks et al., 1998).

There are several important parameters that control economic feasibility of phytomining (Sheoran et al., 2009):

- metal price,
- plant biomass yield,
- the highest achievable metal content in plant.

The last two parameters are important in phytoextraction as well. There are two basic strategies in phytomining (Brooks et al., 1998). The first one suggests using of natural hyperaccumulators of specific metal producing high biomass. It is a cheap method but rarely hyperaccumulators producing high biomass are found (Shah, Nongkynrih, 2007). The second strategy uses chemical complexing agents added into soil to improve metal accumulation and is more expensive. There is also a possibility to produce a "bio-ore" like a by-product of phytoremediation.

Table 2. Plant species used in phytomining of selected metals

Metal	Plant Species	Reference
Ni	*Alyssum bertolonii*	Robinson et al., 1997
	Berkheya coddii	Robinson et al., 1999
	Streptanthus polygaloides	Brooks et al., 1998
	Hybanthus floribundus	Sheoran et al., 2009
	Alyssum murale	Li et al., 2003 Abou-Shanab et al., 2006
	Alyssum corsicum	Li et al., 2003
Co	*Berkheya coddii*	Robinson et al., 1999
Tl	*Iberis intermedia*	Sheoran et al., 2009 Scheckel et al., 2007
	Biscutella laevigata	Sheoran et al., 2009
Au	*Brassica juncea*	Lamb et al., 2001 Bali et al., 2010
	Berkheya coddii	Lamb et al., 2001
	Medicago sativa	Bali et al., 2010

From the economic point of view this method is more suitable for metals with high price in the market. It was calculated that making the phytomining viable for example for a gold recovery would require a crop harvest dry biomass of 10 t from 1 ha with gold concentration of 100 mg.kg^{-1}. The yield would be 1 kg of gold per hectare (Anderson et al., 2005). The use of plants for nanoparticle formation is a perspective process. It was found that nanoparticles of gold, silver and copper can be produced by plants (Haverkamp et al., 2007, Harris, Bali, 2008, Haverkamp, Marshall, 2009).

Phytomining was used for recovery of several metals, mostly nickel, gold, silver and platinum-group metals. Table 2 shows the list of plants used in phytomining.

Phytomining Field Case Studies

Fazenda Brasileiro Gold Mine, Brazil

Gold in the studied site occurred in tabular form as small particles filling cavities or boxworks and was generally associated with arsenates and sulfates, less with hematite and goethite. Gold concentration in the ore was approximately 0.65 g.t^{-1}. The research was conducted in a mini ore pad at the top of a waste-rock dump at the mine site. Two plants *Brassica juncea* and *Zea mays* were used for gold phytoextraction/phytomining. Six weeks after seeding the plot was treated with cyanide and a combination of thiocyanate and hydrogen peroxide to induce uptake of gold. The highest gold recovery during a single treatment cycle was 18% by *Brassica juncea* (39 mg.kg^{-1} of gold in dry biomass weight) planted on the ore treated with NaCN which means that each plot would require about 3 – 4 cycles to recover the same percentage of gold as it occurs from a typical heap leaching operation. To obtain gold concentration in plant 100 mg.kg^{-1} gold concentration in waste ore should be 2 g.t^{-1}. Although according to this research gold phytoreclamation would not be competitive with conventional recovery processes, it shows the perspective especially for the reclamation of a spent heap-leach pile or for a small tonnage ore deposit (Anderson et al., 2005).

Pojske, Pogradec, Albania

The study area is a wide ultramafic site with native *Alyssum murale*. The experimental site was covered by spontaneous native ultramafic vegetation before the beginning of the experiment. During experiment soil was fertilized with fertilizers containing N, P, K and S. The total Ni concentration in the soil was about 3500 mg.kg^{-1}, exchangeable Ni concentration of Ni in the soil was more than 100 mg.kg^{-1}. Except of the *A. murale* the most frequent species were *Chrysopogon gryllus* and *Trifolium nigrescens*. *A. murale* as a Ni hyperaccumulator accumulated 9129 mg.kg^{-1} of Ni. During one year these three plants together accumulated 24.93 kg.ha^{-1} of Ni with the highest amount (22.6 kg.ha^{-1}) extracted by *A. murale*. Ni accumulation by *A. murale* was only little affected by fertilization but its biomass production was significantly higher after fertilization resulting in higher Ni extraction from soil. According to authors of the study total income per ha would be 750 US dollars (calculated for the prices of Ni valid at the time of the study) (Bani et al., 2007).

Mont Pelato, Livorno, Italy

Native plant *Alyssum bertolonii* was selected for the Ni phytomining study lasting 2 years. During the study fertilizers were used to induce higher biomass production. Concentration of Ni in soil was 0.16%. Dry biomass Ni content was 0.69% and 0.72% for vegetative and reproductive tissues, respectively. Ash from the biomass contained about 10% of Ni. Addition of N, P, K fertilizer increased annual biomass production by about 300%. Approximately 72 kg.ha^{-1} of nickel accumulation was calculated (Robinson et al., 1997).

Chapter 2

STRESS PHYSIOLOGY

In biological terms it is difficult to define stress because a condition which may appear the stress for one plant may be optimum for another one (Otte, 2001). The most practical definition of a biological stress is an adverse force or a condition which inhibits the normal functioning and well being of a biological system such as a plant (Mahajan, Tuteja, 2005). Usually the term "stress" is used to indicate the state when a plant is under the influence of stressors. Stressors (stress factors) represent different negative effects of the environment which seriously endanger plants. They can inhibit vital functions of plants, damage their organs or finally lead to the plant death (Procházka et al., 1998). However, rarely only one factor influences plants. More often several factors interact together; some of them even separately do not have to represent stressors for plants (Slováková, Mistrík, 2007). For example, plants in metal-enriched soils frequently suffer from the drought stress mainly because of poor physical soil conditions and a shallow root system (Taulavuori et al., 2005). In the case of plants the stress is complicated by the fact that plants cannot escape from the stress by changing the place so they were forced to evolve a series of molecular responses to cope with stressors. The plant response to the stress is not steady or simple; it is rather a dynamic complex many reactions creating of a comprehensive network of signaling pathways (Shao et al., 2007b).

In general, stress factors can be divided into two main groups – abiotic and biotic. Stress that results from changes in physical features of the environment is often qualified as an environmental stress (Qureshi et al., 2007). Abiotic and biotic stressors are usually divided into several categories (Fig. 2).

As plants are sessile organisms, they cannot avoid being exposed to unfavorable environmental conditions so that they have to use other ways to defense themselves. In principle, two basic responses can be distinguished – **stress avoidance** and **stress tolerance** (Procházka et al., 1998). Stress avoidance includes several forms – inhibition or reduction of elements, or compounds penetration into the plant organism due to various protective structures, avoiding unfavorable conditions (for example by deep rooting or leaf modification) or changes in life cycle (for example growing only in the time of favorable conditions, changes in the time of flowering) (Grime, 2001, Slováková, Mistrík, 2007). However, this way of the protection/defense has only a passive and long-term nature. But in the case that these mechanisms are not efficient against particular stressor, a plant has to use mechanisms of active stress tolerance which reduce negative impact of stressors when they enter cells (Procházka et al., 1998).

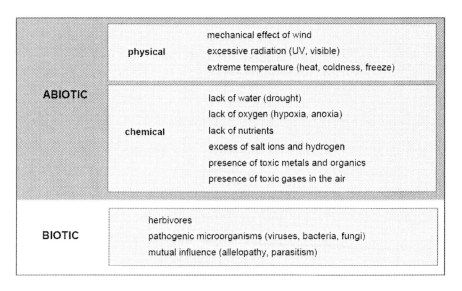

Figure 2. Abiotic and biotic stressors in plant physiology.

Stress is first perceived by the receptors present on the membrane of the plant cell, the signal is then transduced by signal molecules (abscise acid – ABA, jasmonic acid, ethylene, calcium) which spread the information at the cell level within the whole plant (Grichko, Glick, 2001, Mahajan, Tuteja, 2005) resulting in changes in gene expression and generation of compounds what finally leads to changes in both plant development and function (Durand et al., 2010). This chain reaction is called stress reaction (Fig. 3).

Stress reaction may result in changes in plant genetic information (*adaptation*), non-inheritable physiological changes (*acclimation*), or in the case that plant cannot overcome unfavorable environmental conditions inhibition of vital function finally leads to the *death* of the cell and even the organisms (Slováková, Mistrík, 2007, Jansen et al., 2008).

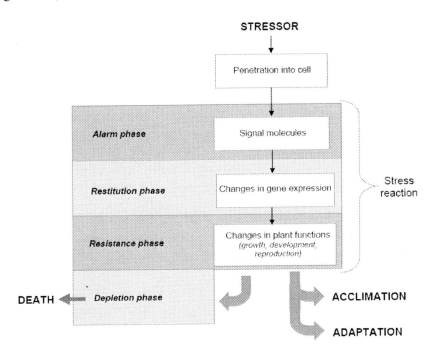

Figure 3. Scheme of plant stress reaction.

Acclimation is often already induced by doses that do not induce "visible" stress symptoms (Potters et al., 2010). Stress reaction can be also divided into four phases:

1. alarm phase (immediately at the beginning of the stress, cell structures and their functions are disrupted),
2. restitution phase (mobilization of compensating mechanisms, it starts when stressors are not lethal),
3. resistance phase (increase of resistance against acting stressors),
4. depletion phase (starts after a long-term and intensive stress duration).

However, such course of stress reaction is rather simplified because when describing stress it is necessary to take into account variety of stressors acting together, complexity of plant response from molecular level up to level of the whole plant. Genetic dispositions of plant (adaptation capability) as well as intensity and duration of stress are also important (Prochádzka et al., 1998).

The increase of plant resistance is usually accompanied with additional energetic costs mainly to metabolites synthesis. Some other metabolism changes ensuring high resistance (for example maintaining of increased concentration of osmolytes) are followed by the decrease of new energy and material sources acquisition rate and thus by the decrease of biomass production rate. Changes in the structure and functions of plants can persist long time after termination of the stress. On the other hand, stress (for example low temperature) can lead to activation of many important morphogenetic processes such as germination or flowering (Brun et al., 2003).

The study of stress in plants growing in natural conditions is complicated by the fact that usually more stressors are acting together. Their mutual interactions can significantly change the character of stress reaction in comparison with the conditions when each stressor acts separately (Prochádzka et al., 1998). The response of plants to a combination of two different abiotic stresses is unique and cannot be directly extrapolated from the responses of plants to each of the corresponding stresses applied individually (Shao et al., 2007a).

Despite many differences it is possible to find several similarities in acclimation responses of plants growing in stress conditions. The most common metabolic changes in the plant cell during stress (Fujita et al., 2006, Bajguz, Hayat, 2009) are as follows:

- generation of stress proteins
- generation and removal of reactive oxygen species
- generation of "stress" hormones
- accumulation of osmolytes and osmotic adjustment

GENERATION OF STRESS PROTEINS

Reactions of plants to abiotic stresses are generally expressed by both regulation and reduction of several protective proteins synthesis such as chaperones and enzymes (Qureshi et al., 2007). During first minutes dramatic changes in quantitative and qualitative presence of proteins in cells occur. The production of some of them sharply increases, while production of others decreases. Proteins which are not present in a cell in normal conditions are formed. These changes culminate within first hours from the beginning of the stress and then slowly normal state returns. Some of stress proteins production is connected with the particular stress, but some of them are produced non-specifically under various stresses. The proteins which are produced regularly during the stress reaction belong to three groups:

1. molecular chaperones
2. proteases
3. ubiquitine

Chaperones regulate changes in protein conformation during membrane transport but they are also able to repair slightly damaged protein conformation. The protein which is damaged in great extent and cannot be repaired is marked by ubiquitine and destroyed by proteases into aminoacids (Procházka et al., 1998, Palma et al., 2002). Except of above mentioned proteins there are also other ones produced during stress but they are rather specific for the particular stress such as proteins induced by high or low temperature, dehydration etc. (Masarovičová et al., 2008).

GENERATION AND REMOVAL OF REACTIVE OXYGEN SPECIES

The common stress response to stress conditions is the excessive production of reactive oxygen species – ROS (Potters et al., 2010). ROS are produced in plants also during normal aerobic metabolism when electrons from electron transport chains are leaked and react with O_2 in the absence of other acceptors (Ashraf, Harris, 2004) but during stress their generation rises significantly. Under optimal conditions both their production and removal are in balance. Oxidative damage occurs when capacity of cell antioxidant systems is exhausted. To protect against ROS action plants have adopted several defense systems responsible for active removal of these radicals (Lee et al., 2007). Generally these systems can be divided into two groups – enzymatic and non-enzymatic antioxidant systems (Shao et al., 2008a). Some plants have even very efficient ways how to avoid ROS generation. For example cowpea (*Vigna sp.*) excretes from roots siderophores which binds metals in the immediate vicinity of the roots. This process restrains metal ions to enter the cell resulting in apparent decrease of free radical formation (Dimkpa et al., 2009). Plants containing high concentrations of antioxidants show considerable resistance to the oxidative damage cause by ROS (Ashraf, Harris, 2004). As ROS production is one of the main responses of plants to heavy metal-induced stress that issue is described in more detail in the chapter 3 Metal stress.

GENERATION OF "STRESS" HORMONES

The production of "stress" phytohormones, such as abscise acid, jasmonic acid, methyljasmonate, polyamines, also belongs among general plant responses to the environmental stress. Besides stress physiology, phytohormones are mainly involved in plant development (Schüler et al., 2004). Abscise acid (ABA) is one of the most important phytohormones in plant stress regulation. ABA is defined as a stress hormone because of its rapid accumulation in response to stresses and its mediation of many stress responses that help plant survival over the stresses (Zhang et al., 2006). For example, dehydration triggers the biosynthesis of the ABA phytohormone which reduces negative impact of water deficit by the stomata closing, increasing of root conductivity etc. (Shao et al., 2007a, Masarovičová et al., 2008). The application of ABA to a plant mimics the effect of stress conditions (Mahajan, Tuteja, 2005). ABA is in large part involved in plant acclimation/adaptation process to many different stresses (Xiong et al., 2002). Other hormone, salicylic acid is thought to amplify ROS signals in a feedback reinforcement loop. Both, salicylic and jasmonic acids play an important role in signaling of biotic stresses (Foyer, Noctor, 2009). Salicylic acid is also known to play an important role in plant protection against heavy metal effects (Krantev et al., 2008).

ACCUMULATION OF OSMOLYTES

Osmotic adjustment to stress can occur by the accumulation of high concentrations of either inorganic ions or low molecular weight organic compounds in the intracellular environment (López-Bucio et al., 2000, Kumar, 2009) Organic osmolytes, unlike the common inorganic ions, do not perturb cell macromolecules so they are generally called compatible osmolytes or solutes (Mehrabi et al., 2008). Except of the osmotic adjustment they act as osmoprotections which includes protection of thylakoid and plasma membrane integrity, stabilizing proteins, acting as a sink for energy or reducing power, being a source of carbon and nitrogen for recovery and scavenging ROS that are by-products of stress (Parvanova et al., 2004, Lee et al., 2008). This mechanism is typical for plants under salinity stress, heat or cold stress and water stress (Prochádzka et al., 1998). The compatible osmolytes generally found in higher plants are low molecular weight sugars, organic acids, polyols and nitrogen containing compounds (aminoacids, amides iminoacids ectoine, proteins and quaternary ammonium compounds (Ashraf, Harris, 2004).

Chapter 3

METAL STRESS

Heavy metal contamination is one the most serious environmental problems limiting plant productivity and threatening human health (Gambale, 2001). The group of heavy metals involves approximately 40 elements with the density higher than 5 $g.cm^{-3}$, some authors define heavy metals as the group of elements with the atomic number higher than 33. Usually metalloids arsenic and selenium are also included among heavy metals because of their similar characteristics (Gažo et al., 1974). In according to Duffus (2002) the term heavy metals appears to be commonly applied to elements of density higher than $3.5 - 7$ $g.cm^{-3}$ and atomic number higher than 20 and includes transition metals, some metalloids, lanthanides and actinides.

Inorganic ions present in the environment have several ways how to influence the organisms. Their toxic effects depend on the kind of metal (e.g. the phytotoxicity of some relatively common heavy metals such as Cd, Cu, Hg and Ni is substantially greater than that of Pb and Zn), their speciation in the environment and their biological availability (Raskin et al., 1994, Escuder-Gilabert et al, 2001).

3.1. SOURCES OF METALS IN SOILS

Soils contain a repertoire of diverse heavy and trace metals. They are natural components of soil at trace levels but human activities have contributed to their undesirable accumulations in the environment (Kabata-Pendias, Pendias, 1992). There are many sources from which high concentrations of metals can originate. These sources can be generally categorized as natural and anthropogenic.

Soils with naturally higher concentrations of heavy metals such as Zn, Pb, Ni, Co, Cr, Cu and Mn, Mg, Cd, Se occur on parent rocks which are usually considered ores. Some of the heavy metals are main constituents of these ores; some of them represent trace components of minerals in ores (Gardea-Torresdey et al., 2005). They enter crystal lattice of these minerals by isomorphic exchange of main mineral elements. Igneous ores usually contain higher concentrations of heavy metals than sedimentary, but sedimentary ores are more abundant, covering approximately 75% of Earth surface representing more important source of heavy metals in soils (Bláha et al., 2003)

Agriculture, industry, cities and transportation are main sources of anthropogenic pollution of soils. Industrial sources include mining activities, metallurgy, energetics, electronics and pigments. Agricultural sources of metal pollution include both inorganic and organic fertilizers, liming application to raise soil pH, sewage sludge, irrigation waters and pesticides (Freedman, 1989, Chen et al, 2000, Orcutt, Nilsen, 2000). Waste – municipal as well as industrial is a new big source of pollution. Dumps and landfills contaminate environment because of water and wind erosion. Industrial and municipal sources are the main sources of metals in the environment.

Al and Fe are two most abundant metals in soil (Feng, 2005). Lead is the principal contaminant in soil and waste deposits; however, other metals including cadmium are typically found in lower concentrations (Pichtel et al, 2000) but even in these low concentrations they are toxic to plants.

Pb is naturally found deep within the earth and mined together with silver deposits. It is the most common and abundant contaminant of the environment and human body. Sources of Pb contamination originate from mining and smelting of lead-ores, burning of coal, effluents from storage battery industries, automobile exhausts and automobile demolition, metal plating and finishing operations recycling of metals, fertilizers, pesticides and from additives in pigments and gasoline (Freedman, 1989, Pichtel et al, 2000, Verma, Dubey, 2003).

Cd is mined as part of zinc deposits. It is less abundant than Pb but it has entered the air, food and water thus making it a major environmental concern. Cadmium is one of the most toxic metals in the present environment and it is more toxic to humans than lead (Cho, Seo, 2004). The sources of cadmium are mining activities, metal smelters, coal burning power stations, sewage sludge, leachates from landfills (Chen et al, 2000) as well as application of Cd-containing pesticides and phosphate fertilizers (Feng et al., 2010).

Higher content of Cu in the environment results not only from its increased use in industrial activities like mining and smelting, but also from its use as pesticide and its presence in sewage sludge amendments (Ke, 2007).

Cr contamination can originate from natural sources but it mainly comes from several industrial and agricultural activities such as ore refining, electroplating, phosphate fertilizers and waste disposal on land (Vernay et al., 2007). Big amount of chromium originates from leather processing and finishing, production of refractory steel, drilling muds, catalytic manufacture, wood preservation, cooling tower water treatment and production of chromic acid (Shanker et al., 2005).

Ni in soil originates naturally from ultramafic and serpentinic soils as well as from discharged municipal and industrial sewage (Drążkiewicz, Baszyński, 2010).

Main sources of Zn soil contamination are sewage sludges or urban composts, fertilizers, emissions from municipal waste incinerators, residues from metalliferous mining and metal smelting industry (Yadav, 2010).

Heavy metals can be also transported for long distances contaminating areas far away from the source of their production (Taulavuori et al., 2005). For example, it has been calculated that 2 - 6% of certain metal pollution (Sb, Cd, As, V, Zn and Pb) from Eurasia is transported to the North Pole (Akeredolu et al., 1994).

Soil metal contamination is rarely caused by only one metal but more often several toxic metals are present together making the situation for plants even more complicated (Murakami, Ae, 2009).

3.2. METAL PHYTOTOXICITY SYMPTOMS

Phytotoxicity is mainly associated with non-essential metals like As, Cd, Cr and Pb (Clemens, 2006b). When metal ion enters the cell, several enzymes and redox systems are activated. The first symptom of metal toxic effect is the inhibition of cell division and root prolongation. Most of the metals are accumulated and stored mainly in plant roots. Part of the metals is translocated to upper parts where it influences the physiological processes in leaves, mostly photosynthesis (Procházka et al., 1998).

Metal accumulated in cells can differently interact at both the cell and molecular levels damaging plant and causing various visible toxicity symptoms. The influence of these metals depends on plant growth phase – in young plants inhibition of growth occurs most often, in old plants senescence is usually accelerated (Ryser, Sauder, 2006). Young plants are generally more sensitive to the presence of metals then older ones (Slováková, Mistrík, 2007). Metals may also influence the allocation of sexual reproduction and delay flowering (Brun et al., 2003). Flowering phenology is one of the sensitive variables in response to soil metals (Ryser, Sauder, 2006).

Excessive levels of metal ions in plants can cause various stress responses of parallel and/or consecutive events manifested outwardly by visible symptoms. The most common toxicity symptoms of selected heavy metals are described bellow.

Lead has not been shown to be essential in plant metabolism, although it occurs naturally in all plants and it is a highly toxic element. Responses to lead exposure include the decrease in root elongation and biomass production, inhibition of chlorophyll biosynthesis, inhibition of photosynthesis and respiration by affecting electron transport mechanisms, induction or inhibition of several enzymes as well as cell disturbances and chromosomal lesions (Xiong, 1997, Orcutt, Nilsen, 2000).

Cd is easily taken up by roots and translocated to different plant parts. A high Cd accumulation generally causes growth inhibition and even the plant death due to the changes in enzyme activity, photosynthesis, respiration, transpiration, and nutrient uptake (Cho, Seo, 2004, Jia et al., 2010). Symptoms of Cd toxicity are generally browning of root tips, chlorosis, necrosis, leaf rolling or drying and growth inhibition. Cd toxicity is probably caused by competition of Cd with Zn for binding sites in biomolecules and also substitution of Ca in the photosystem II reaction centre (Faller et al., 2005, Krantev et al., 2008, Durand et al., 2010).

Al, although it does not belong among heavy metals, has the similar impact on plants as heavy metals. It is together with oxygen and silicium one of the most abundant elements in the Eart crust, its content is approximately 7%. For this reason it is potentially highly toxic to ecosystems. First visual symptoms of Al toxicity is inhibition of root prolongation. In addition Al toxicity symptoms include reduction in biomass production, leaf chlorosis, reddening of leaf veins, yellowing and drying of leaf tips which are in fact symptoms of lack of phosphorus and calcium (Al form insoluble $AlPO_4$ and blocks Ca-channels) (Slováková, Mistrík, 2007).

Cu is an essential element for plants at low concentrations but at high concentrations it is toxic (Hattab et al., 2009). Cu is reported to modify a number of physiological processes and particularly chlorophyll degradation (Prasad et al., 2001). The most often general chlorosis and stunting of growth are observed at even slight excess of Cu (Farooqui et al., 1995) as well as browning of leaves and leaf loss (Macinnis-Ng, Ralph, 2002).

Uranium causes chlorosis because of blocking iron accumulation by roots (Viehweger, Geipel, 2010).

Chromium usually occurs in two forms as Cr^{6+} and Cr^{3+}. Cr^{6+} is known to be more toxic to living organisms due to its ability to pass the membrane, penetrate the cytoplasm and react with the intracellular material. Chromium interferes with several metabolic processes causing inhibition of seed germination, reduction of root elongation, growth and yield reduction, foliar chlorosis, stunting, wilting and finally the plant death (Shanker et al., 2005, Vernay et al., 2007).

Excess of arsenic in soils inhibits seed germination and plant growth, decreases plant height, tillering, fruit and grain yield and finally it might lead to the plant death (Azizur Rahman, et al., 2007).

The zinc presence in the soil reduces Ni toxicity and seed germination, increases ATP/chlorophyll ratio (Gardea-Torresdey et al., 2005) and leads to significant reduction in Fe concentration in plants (Pavlíková et al., 2008). High levels of Zn in soil inhibit plant growth and cause senescence, chlorosis in the young and eventually also old leaves (Yadav, 2010).

Growth reduction, chlorosis, alternations in water relations, gas exchange, photosynthetic activity and enzymatic activity are symptoms of Ni toxicity. Ni is known to trigger reactive oxygen species. Ni causes Fe, Mg and Ca deficiencies (Drążkiewicz, Baszyński, 2010).

Toxicity of Hg is usually expressed by the decrease of photosynthetic activity, water uptake and antioxidant enzymes (Gardea-Torresdey et al., 2005). High level of Hg interfere the mitochondrial activity and induces oxidative stress (Yadav, 2010).

3.3. MECHANISMS OF METAL ACCUMULATION BY PLANTS

The metal accumulation in various parts of the plant depends upon availability and the species of metals in soils, solubility, their translocation potential, and the type of plant species (Lasat, 2002, Sinha et al., 2009). Molecular understanding of plant metal accumulation leads to better understanding of phytoremediation and widening of its applications. Except of cleaning up environment the phenomenon of metal accumulation in plants can be used also in other fields for example in human nutrition. Engineering Zn accumulation in edible plant parts might help in enriching diets for Zn. Foliar application of Se increases the antioxidant activity of tea and supplies human body with Se which is necessary for the protection against several diseases and cancer as well (Xu et al, 2003). Conversely, most of the toxic non-essential elements such as Cd enter the human body via plant-derived material (i.e. food or tobacco smoke). Identification of mechanisms governing the metal accumulation process could result in development of plants with lower metal content (Clemens et al., 2002).

Ability of organisms to survive depends besides other factors, on their abilities to cope with toxic properties of various materials present in their environment. Organisms have adapted to higher toxicity in environment by evolving mechanisms to maintain low intracellular concentrations of toxic pollutants. On the other hand, many of the elements, including metals, are as trace elements essential for normal development of organisms. They play an important role in organisms. Metals which may not be as yet identifiable as serving a beneficial biologic function are referred to as nonessential. Certain concentration of essential metals is necessary for optimal growth of organisms, but an oversupply results in toxic effects and lethality in the end. Nonetheless, there is no doubt that all metals are potentially hazardous to living organisms, and not necessarily at large exposure levels (Förstner, Wittmann, 1979). The concentration of essential elements in the environment is usually lower than requirement of organisms, so they had to develop mechanisms able to sequestrate and concentrate these elements from the environment (Wood, Wang, 1985). But because control of accumulation is imperfect, organisms have to cope with exposure to unwanted elements. A lack of specificity of uptake and distribution systems also leads to the accumulation of metals such as Cd, As or Sb, which is generally considered nonessential (Wood, Wang, 1985, Clemens et al., 2002).

Metal accumulation by plants can be divided into three steps:

1. mobilization, root uptake and sequestration,
2. translocation,
3. tissue distribution and storage.

1. Mobilization, Root Uptake and Sequestration

A very important step for accumulation of metals is their sequestration by roots as growth media of plants – nutrient solutions, soil etc. are the main sources of metals. Kabata-Pendias (2001) distinguished three basic processes responsible for metal uptake by plant roots:

1. cation exchange by roots,
2. transport inside cells by chelating agents or other carriers,
3. rhizosphere effects.

The initial contact a metal has with a plant is usually in roots and at the membrane level, although heavy atmospheric deposition of elements can cause the effect to be initiated in the shoots (Clemens et al., 2002). Uptake of metals is mainly influenced by their bioavailable fraction rather than by the total amount in soil (Vamerali et al., 2010). The actual metal bioavailability depends on the soil and metal physical - chemical properties as well as metal content, water content and other elements in the rhizosphere (Yang et al., 2005) and for some metals it can be limited because of low solubility and strong binding to soil particles (Clemens, 2006a). For example, iron is mainly present in the form of insoluble hydroxides, but availability of Zn is less restricted, whereas the bioavailability of some of the target metals in phytoremediation, particularly Pb, is limited. Low soil availability of metals can be a major factor limiting the application of phytoremediation. But plants can actively contribute to metal availability. They enhance the metal accumulation by two basic mechanisms – acidification of the rhizosphere and the exudation of carboxylates (Zhao et al., 2001, Clemens et al., 2002). Acidification of rhizosphere soil was found in several plant species accumulating Cu, Ni, Zn and Cd (Yang et al., 2005). Root exudates of plants are composed mainly of amino acids (e.g., aspartic, glutamic, prolinic) and vary with plant species (and varieties), microorganism association, and conditions of plant growth (Kabata-Pendias, 2001). For example, species from the family *Poaceae* or *Fabaceae* secrete phytosiderophores (chelating agents) to solubilize soil Fe and accumulate the intact chelate into root cells (Chaney et al., 1997, Dimkpa et al., 2009). Roots can reduce soil-metal ions by specify plasma membrane bound metal reductases, e.g. pea plants deficient in Fe and Cu have an increased ability to reduce Fe^{III} an Cu^{II} that is coupled with an increased uptake of Cu, Mn, Fe and Mg (Raskin et al., 1994). Iron accumulation by roots of dicotyledonous plants is based upon reduction of exogenous ferric iron to ferrous iron by reductases and the subsequent transport across the root plasma membrane by Fe^{II} transporters. Similar carrier systems transport a broad range of divalent cations (Cu^{2+}, Ni^{2+}, Zn^{2+} and Cd^{2+}) into plant root cells (Briat, Lebrun, 1999). Besides it, root-colonizing bacteria, as well as mycorrhiza, have a large impact on the availability of heavy metals for plant uptake (Fitz, Wenzel, 2002, Glick, 2003). For instance, soil bacteria significantly enhance Se and Hg accumulation by the saltmarsh bulrush (*Scirpus robustus*) and rabbit-foot grass (*Polypogon monspeliensis*) (de Souza et al., 1999).

Application of chelators such as ethylenedimanetetraacetic acid (EDTA), ethyleneglycoltetraacetic acid (EGTA), ethylenediaminedisuccinic acid (EDDS) to the soil to enhance metal bioavailability, especially Pb, was studied (Wu at al., 1999, Barona et al., 2001, Kos, Leštan, 2003, Hornik et al., 2009), but the risk of this technique is that metal-chelate complexes are very soluble and leach into the groundwater easily (Römkens et al., 2002, Madrid et al., 2003). This treatment can have in same cases also negative effect on soil and plant health. The application of chelators was found sometimes to cause foliar necrosis, transpiration reduction, biomass production, reduction of soil micro-fauna etc. (McGrath et al., 2006).

Substances pass into roots through cuticle-free unsuberized cell walls. Therefore, roots absorb substances far less selectively than leaves. Environmental contaminants enter the roots together with water, like nutrients (Kvesitadze et al., 2006). Transport systems responsible for ion accumulation by roots are not very selective so many metals are accumulated because of their similarity with essential nutrients. For example AsO_4^{3-} is taken up by plant roots on the basis of their similar physico-chemical properties with PO_4^{3-} ions (Beceiro-González et al, 2000). Accumulation of arsenic ions as well as cadmium and antimony into the cell is non-selective and takes place via normal transport systems by chance (Wood, Wang, 1985, Elbaz-Poulichet et al, 2000). Solubilized metal ions may enter the root either via the extracellular (apoplastic) or via intracellular (symplastic) pathways. Apoplastic binding of metallic cations such as copper and zinc can contribute significantly to the total cation content of the roots and may also serve as a metal storage pool (Briat, Lebrun, 1999). Apoplastic transport is limited by the high cation exchange capacity of the cell walls, unless the metal ion is transported as a non-cationic metal chelate (Raskin et al., 1994). Cation uptake selectivity from the soil solution depends upon specific sites located in the plasma membrane of individual cells. For example, surface chemical properties of root cell wall, depending on its $RCOO^-/RCOOH$ composition can account for the different manganese and copper toxicity tolerance exhibited by two tobacco genotypes (Briat, Lebrun, 1999).

Symplastic transport requires that metal ions move across the plasma membrane, which usually has a large negative resting potential. This negative potential provides a strong electrochemical gradient for the inward movement of metal ions (Raskin et al., 1994). Transport systems and intracellular high-affinity binding sites then mediate and drive uptake across the plasma membrane. Several organic acids have been identified as positive reagents to accelerate the absorption of heavy metals by roots (Wu et al., 2010). In roots of the grass *Deschampsia caespitose*, citric acid concentration increased when Zn-tolerant ecotypes were exposed to Zn excess (Rauser, 1999).

From yeast studies, it is apparent that most of the cation transporters show a rather broad substrate range, enabling even non-essential metals such as Cd to enter cells. However, for phytoremediation, specificity and affinity of transport systems have to be considered in relation to relative abundance of different substrates. Non-target elements such as Ca might out-compete target elements (Clemens et al., 2002). Even externally added chelates enhance desorption of some metals (e.g. Pb) from soil to soil solution, facilitate their transport into xylem and increase their translocation from roots to shoots (Barona et al., 2001).

2. Translocation

Metals absorbed by roots are translocated to different plant organs by the same physiological processes as those used to transport nutrients (Kvesitadze et al., 2006). In general, long-distance transport of metals in higher plants is carried out by the vascular tissues (xylem and phloem) and is partly related to the transpiration (Kabata-Pendias, 2001).

The apoplast continuum of the root epidermis and cortex is readily permeable for solutes. The cell walls of the endodermal cell layer act as a barrier for apoplastic diffusion into the vascular system. Solutes have to be taken up into the root symplasm before entering xylem. Xylem-loading system controls metal accumulation from external solutions to xylem stream. This process is mediated by membrane transport proteins (Cabañero, Carvajal, 2007, Mori et al., 2009).

Chelatation with some ligands probably determines the transport of metals. For example nicotianamine is necessary for the redistribution of Fe, Zn, and Mn via the phloem and is required for Cu transport in the xylem. Takahashi et al. (2003) found that it may also be required for the intracellular regulation of metal binding proteins, such as Zn-finger proteins. Chelation with other ligands, for example histidine and citrate, appears to route metals primarily to the xylem (Clemens et al., 2002). Organic acids, especially citrate, and amino acids are the main metal chelators for some metals such as Fe, Pu, Ni and Cd in xylem. In the case of nickel, free histidine increasing in the xylem sap enhanced translocation of this metal to the shoots, and could explain nickel hyperaccumulation in some plants (Briat, Lebrun, 1999). By contrast, chelation with other ligands, such as phytochelatins or metallothioneins, might route metals predominantly to root sequestration (Clemens et al., 2002).

Xylem-unloading processes represent the first step in controlled distribution and detoxification of metals in the shoot (Raskin et al., 1994, Clemens et al., 2002). Except of the xylem transport which predominantly transports metal ions and water, part of the metals can be in the form of assimilates re-distributed from leaves to the parts of a plant below (shoot axes, roots) and above (shoot tops, fruits) the leaves, passing through sieve tubes in the phloem (Kvesitadze et al., 2006). Metal transport among plant organs depends on the kind of metal. In general, Ag, B, Li, Mo, and Se are easily transported from roots to above-earth parts; Mn, Ni, Cd, and Zn are moderately mobile; and Co, Cu, Cr, Pb, Hg, and Fe are strongly bound in root cells (Kabata-Pendias, 2001).

3. Tissue Distribution and Storage

Metals reach the apoplast of leaves in the xylem sap, from where they have to be scavenged by leaf cells. Transporters mediate uptake into the symplast, and distribution within the leaf occurs via the apoplast or the symplast. Metal sequestering occurs inside every plant cell, maintaining the concentrations within the specific physiological ranges in each organelle and ensuring delivery of metals to metal-requiring proteins. The excess of heavy metals is sequestered in leaf cell vacuoles (Boojar, Goodarzi, 2007, Mleczek et al., 2009). The distribution pattern varies with plant species and element. Furthermore, trichomes apparently play a major role in storage and detoxification of metals (Clemens et al., 2002, Broadhurst et al., 2004).

Intracellular binding and sequestration drive the passage of transition metal ions across the plasma membrane. Several processes known to contribute to metal tolerance are associated with metal accumulation at the same time. The exposure to toxic heavy metals or to high concentration of micronutrients induces the synthesis of phytochelatins as one of the principal responses of plants, animals and many fungi (Winge et al., 1998, Malmström, Leckner, 1998, Bang, Pazirandeh, 1999, Gardea-Torresdey et al., 2004). Phytochelatins serve as a mechanism for sequestration of heavy metals in the plant vacuole and present an example of biomineralization (Buchanan et al., 2000). Phytochelatin deficiency is correlated with Cd, Cu and As hypersensitivity. For example, Zhang et al. (2005a) have found that treatment with Cd and As resulted in a strong increase phytochelatins contents in roots of garlic seedlings.

Other potential mediators of metal sequestration and accumulation include members of the Cation Diffusion Facilitator family (CDF). In eukaryotic systems, they have been implicated in moving Zn, Cd and Co from the cytosol into cellular compartments. Studies on various hyperaccumulators and their respective close relatives show a possible correlation between differences in either expression levels or substrate specificities of CDFs and the hyperaccumulation phenotype (Williams et al., 2000, Clemens et al., 2002, Blaudez et al., 2003). Also other classes of proteins are probably connected with metal transport in cells – the heavy metal (or CPx-type) ATPases, the natural resistance associated macrophage protein family of proteins, zinc-iron permease (ZIP) family proteins etc. (Yang et al., 2005) and ferritins (Briat, Lebrun, 1999).

3.4. PLANT ADAPTATION TO METAL STRESS

In the case of toxic metals presence in the environment, plants do not have any possibility to leave so they are forced to develop relevant mechanisms for adaptation to metal stress. In general, two responses can be distinguished in the presence of metals - metal sensitivity and metal resistance. Sensitivity to metals results in injury or death of plants. Resistance means that in spite of metal toxicity a plant responses in a way enabling it to survive high concentrations of metals, and to produce the next generation of plants. Resistance includes avoidance, which describes the mechanisms of external protection of the plant from metal stress and tolerance, which makes the plant be able to survive internal stress imposed by high metal concentrations (Orcutt, Nilsen, 2000). Some plants are even able to accumulate high metal concentrations in their bodies – they are called hyperaccumulators. The content of specific metals in their tissues exceeds levels that are actually required for normal growth and development. For the first time the term "hyperaccumulator" was used by Brooks and his colleagues in 1977 (Brooks et al., 1977) to describe plants able to accumulate high amounts of nickel. Hyperaccumulators can concentrate metals in their aboveground tissues to the levels far exceeding those present in the soil or in the non-accumulating species growing nearby. One of the definitions suggests that a plant containing more than 0.1% of Ni, Co, Cu, Cr and Pb or 1% of Zn in its leaves and stems on the dry weight basis, can be considered a hyperaccumulator, irrespective of the metal concentration in the soil (Raskin et al., 1994). As there are plants growing in natural conditions containing high metal concentrations in their tissues which do not reach concentrations of these metals in surrounding soil, one more requirement has been added into the definition of hyperaccumulator – a plant has to accumulate metal in the amount exceeding its amount in soil (Yang et al., 2005, Gardea-Torresdey et al., 2005). Several plants belong to a group of hyperaccumulators. *Agrostis stolonifera* accumulates 300-times more of arsenic from soil than other plants growing in the same area; *Minuartia verna* contained 1000-times more of cadmium than its amount in soil was (Domažlická, et al., 1994). Plants belonging to the species *Alyssum* and *Thlaspi* are very good nickel hyperaccumulators (Raskin et al., 1994). *Pteris vittata* is a good hyperacumulator of arsenic (Fayiga et al., 2004, Zhang et al., 2004). *Sonchus asper* has hyperaccumulation capacity for Pb and Zn, *Corydalis pterygopetala* has hyperaccumulation capacity for Zn and Cd (Yanqun et al., 2005). *Sedum alfredii* was found to be a hyperaccumulator of Cd (Sun et al., 2007). The list of hyperaccumulators contains many species today and is expanding rapidly. More than 400 plant species are known as hyperaccumulators of various metals, 317 species from them are Ni hyperaccumulators (Baker et al., 2000, Kazakou et al., 2010).

Plants that can grow on soils that are contaminated with high levels of metals are known as metallophytes (Banásová, 1996, Hronec, 1996, Whiting et al., 2004, Gardea-Torresdey et al., 2005). They have specific biological mechanisms enabling them to tolerate high metal concentrations. Some of them can grow only on metal containing soil – they are absolute metallophytes. Other species may occur on contaminated as well as uncontaminated soils (Nessner Kavamura, Esposito, 2010). They are often metal specific and thus can serve as the indicators of metal presence in soil (Bláha et al., 2003). Several species of *Thlaspi* are the most known metallophytes around the world. These species have been found growing in Cd, Pb, Zn, and Ni metalliferous soils in several European countries (Gardea-Torresdey et al., 2005).

The plants tolerant to metalliferous soils can be divided into three groups according to the metal concentration found in their tissues (Baker, Brooks, 1989):

1. accumulator,
2. indicator,
3. excluder.

The accumulator is a plant able to uptake very high levels of ions to the extent exceeding the levels in the soil. The indicator can take up metals at a linear rate relative to the concentration of metal in the soil. The excluder takes up metals but restricts increased concentrations in the shoots until a critical level is reached, above which metal concentrations start to increase in the shoots. For example, copper exclusion was found as the mechanism responsible for Cu tolerance in *Malva sylvestris* (Boojar, Goodarzi, 2007). Fig. 4 illustrates these three types of tolerant plants.

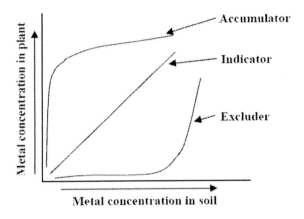

Figure 4. Three types of tolerant plants (adapted from Orcutt, Nilsen, 2000).

Plants differ among species with respect to how much of a specific metal they can accumulate in their shoots. A single plant species may also differ with respect to uptake, transport and accumulation of different metals. The place where the metals are stored varies in different plants species, as well. These differences are usually based on leaf/root ratios for specific elements which are both species specific and metal specific (Orcutt, Nilsen, 2000).

Hyperaccumulation is one of the fundamental characteristics for plant generally used for phytoremediation. Almost all metal-hyperaccumulating species known today were discovered on metalliferous soils, either natural or man-made, often growing together with metal excluders. Actually, almost all metal hyperaccumulating plants are endemic to these soils, suggesting that hyperaccumulation is an important ecophysiological adaptation to heavy metal stress and one of the manifestations of heavy metal resistance (Raskin et al., 1994, Sun et al., 2007, Maestri et al., 2010). The basic characteristics of hyperaccumulating plants are as follows (Chaney et al., 1997):

- a plant must be able to tolerate high levels of the element in both root and shoot cells: hypertolerance is the first property that makes hyperaccumulation possible,
- a plant must have ability to translocate an element from the roots to the shoots at high rates,
- there must be a rapid uptake rate for the element at the level that occur in soil solution.

Uptake and sequestration of toxic materials represent an interesting biological puzzle. The metals hyperaccumulated by plants are generally viewed as toxic in relatively low doses; the logical extension of this principle is that the metal levels in these plants might render them relatively toxic to the organisms they interact with (Boyd, Martens, 1998).

3.5. METAL-INDUCED STRESS

The phytotoxicity symptoms seen in the presence of excessive amounts of heavy metals may be due to a range of interactions at the cellular/molecular level (Hall, 2002). Three different molecular mechanisms of heavy metal toxicity in plants can be outlined according to their distinct chemical and physical properties:

a. stimulated generation of reactive oxygen species (ROS) by autoxidation and Fenton reaction that modify the antioxidant defense and elicit oxidative stress,

b. blocking of essential functional groups in biomolecules, direct interaction with proteins due to their affinities for thiol-, histidyl- and carboxyl-groups, causing binding of metals to target structural, catalytic and transport sites of the cell. For example Hg^{2+} ions inhibit the activities of antioxidant enzymes especially of glutathione reductase, and also raise a transient depletion of GSH,

c. displacement of essential metal cations from specific binding sites, causing functions to collapse. For example, Cd^{2+} replaces Ca^{2+} in the photosystem II reaction centre, causing the inhibition of PSII photoactivation and Hg^{2+}, Cd^{2+}, Cu^{2+} or Ni^{2+} ions replaces Mg^{2+} in chlorophyll (Schützendübel, Polle, 2002, Hall, 2002, Rai et al., 2004, Slováková, Mistrík, 2007, Sharma, Dietz, 2009).

Activated responses to acclimation evoke a cycle of feedback with different sites of heavy metals actions. This process results in the repairing of damaged macromolecules, strengthening of the antioxidant defense system and decreasing of heavy metals content in plasmatic compartments (Sharma, Dietz, 2009). To cope with excessive amount of metal ions in cells, plants have developed diverse mechanisms to maintain and regulate metal homeostasis operating at both intra and extracellular levels (Pavlíková et al., 2008). In general, three areas are involved in the plant response to heavy metal stress in plants (Fig. 5):

1. formation of phytochelatins and metallothioneins and consequent compartmentalization of metals in vacuoles,
2. reduction and consequent stimulation of antioxidant systems,
3. influencing of photosynthetic apparatus.

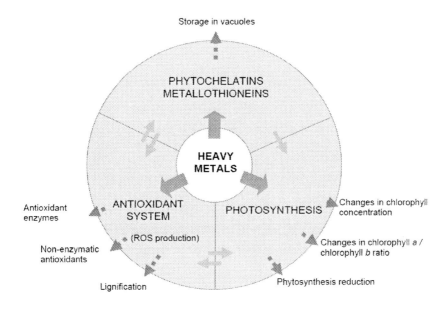

Figure 5. Areas significantly affected by heavy metal-induced stress.

All these three areas interfere mutually and changes in one of them often cause changes in the others. For example, phytochelatins production leads to the decrease of glutathione, which is consumed during the process. But because glutathione is one of the most important molecules with antioxidant activity, its decrease results in the increase of ROS level in the cell. On the other hand, recently has been documented that both phytochelatins and metallothioneins act as scavengers for ROS so they can protect cell metabolism including photosynthesis from oxidative stress (Zhang et al. 2005b) and finally phytochelatin synthesis is regulated by the intracellular level of glutathione which synthesis is regulated by oxidative stress (Hirata et al., 2005). ROS are produced in the stressed cell also in chloroplast by the electron acceptor of photosystem I and in disrupted thylakoids membrane (Hernández et al., 1995). Metals inactivate antioxidant enzymes which directly lead to the increase of ROS. ROS produced in mitochondria because of mitochondrial structures damage influence also respiratory gas exchange. An increased energy demand for active exclusion or sequestration of heavy metals is usually met by increased net respiration rates. Plants exposed to mild stress increase respiratory rates but under strong metal stress respiration rate decreases significantly (Prasad et al., 2001).

3.5.1. Formation of Phytochelatins and Metallothioneins

In the effort to protect themselves against heavy metal toxicity cells had to develop special mechanisms enabling them to regulate the intracellular concentration of metal ions. Synthesis of proteinaceous complexes that chelate and sequestrate metal ions is one of the common plant responses to the presence of heavy metals (Zhang et al., 2005b). The most common metal-binding ligands in plants are phytochelatins. Besides heavy metal detoxification, these compounds are responsible for maintenance of the homeostasis of intracellular level of essential metal ions (Hirata et al., 2005) and a long-distance transport in plants (Clemens, Peršoh, 2009).

Phytochelatins (Fig. 6) are low-molecular-weight thiols rich in cysteine which bind metal ions using the thiol group as the ligand. They consist of only three amino acids – Glu, Cys and Gly and have various sizes with the general structure $(\gamma\text{-Glu-Cys})_n\text{Gly}$ (n = 2 to 11). They are enzymatically synthesized de novo from glutathione by phytochelatin synthase in a very short time after the exposure to toxic metals (Cobbett, 2000, Zhang et al., 2005b, Clemens, 2006a).

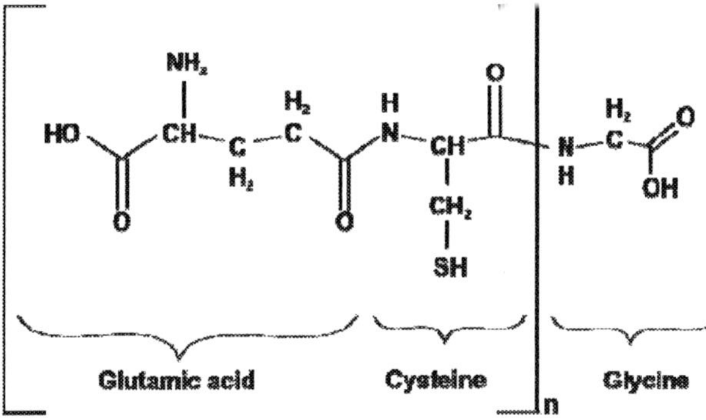

Figure 6. General chemical sturcture of phytochealtins (Bontidean et al., 2003).

Phytochelatin synthase is constitutively present in the cytoplasm of plant cells (Stolt et al., 2003) and it is activated when two glutathione (GSH) molecules with heavy metal form a thiolate and one γ-Glu-Cys moiety is transferred to a GSH free molecule or to a previous synthesized phytochelatin (Mendoza-Cózatl, Moreno-Sánchez, 2006). They are not synthesized not on the ribosomes (Procházka et al., 1998). Phytochelatins loaded with heavy metals are pumped at the expense of ATP into the vacuoles. Because of the acidic environment in the vacuole, the heavy metals are probably liberated from the phytochelatins and finally deposited there (Heldt, 1997, Cobbett, 2000). Process of phytochelatin synthesis is shown in Fig. 7.

The biosynthesis of phytochelatins is induced by heavy metals such as Cd, Hg, Pb, Zn, Cu and Ag. Among these, Cd generally has the highest induction ability and is strongly conjugated by phytochelatins (Tsuji et al., 2002, Hall, 2002). The tendency of metal to induce phytochelatin synthesis for cell suspension culture of *Rauvolfia serpentine* was found to decrease in the order (Grill et al., 1987):

Hg > Cd, As, Fe > Cu, Ni > Sb, Au > Sn, Se, Bi > Pb, Zn

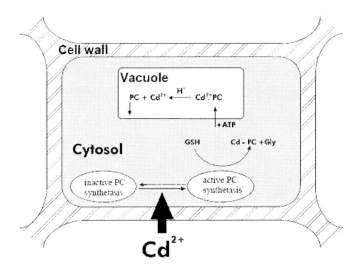

Figure 7. Process of phytochelatin (PC) synthesis in the presence of Cd (adapted from Zenk, 1996).

Although it is generally accepted that phytochelatins play an important role in heavy metal detoxification, there are some studies suggesting that there can be several different mechanisms responsible for the detoxification process especially in metal tolerant species. It is still assumed that phytochelatins have an important role in metal detoxification in metal sensitive plants, but at not all metal tolerant species. For example, the decrease of phytochelatin content in *Silene vulgaris* decreased Cd tolerance only in the sensitive ecotype and had no influence on the tolerance of tolerant ecotype (Persans, Salt, 2000). Wójcik et al. (2005) also showed that phytochelatin production is not the primary mechanism of Cd-tolerance in the Zn/Cd hyperaccumulator *Thlaspi cearulescens*. On the other hand, Figueroa et al. (2008) studying two wild plants chronically exposed to low or moderate heavy metal levels in soil found in one of them (*Ricinus communis*) increased levels of phytochelatins in contrast to the other one *Tithonia diversifolia* with no phytochelatins production. They suggested that induction of phytochelatins might be genotype-dependent.

Increased phytochelatins production has been observed in the presence of various organic chelators (EDTA, EDDS, citric acid) added into nutrient solution or soil (Piechalak et al., 2003, Sun et al., 2005). But this effect can be connected with increased bioavailability of metals in the presence of chelators.

Metallothioneins are also cysteine-rich polypeptides, however, in contrast to phytochelatins, they are products of mRNA translation. They can be induced by metal exposure both in animals and plants (Ohlsson et al., 2008). Metallothioneins typically contain two metal-binding and cysteine-rich domains that give them a dumb-bell conformation (Zhang et al., 2005a). They have been found to be responsible for Cu tolerance in plants e.g. *Arabidopsis* sp. (Murphy, Taiz, 1995). In contrast to metallothioneins phytochelatins have higher metal-binding capacity (on a per cysteine basis) than metallothioneins (Bontidean et al., 2003). Little is known about the relationship between phytochelatins and metallothioneins in plants (Zhang et al., 2005b) it is likely that both phytochelatins and metallothioneins play relatively independent roles in metal detoxification and/or metabolism (Qureshi et al., 2007).

3.5.2. Reduction and Consequent Stimulation of Antioxidant Systems

The generation of reactive oxygen species (ROS) causing oxidative damage to plants is one of the typical heavy metals toxicity symptoms in plants. These ROS include superoxide radical ($O_2^{\bullet-}$), hydroxyl radical ($OH^{\bullet}$) and hydrogen peroxide (H_2O_2) that are produced as by-products during membrane linked electron transport activities as well as by a number of metabolic pathways. ROS are partially reduced forms of molecular oxygen O_2 (Briat, Lebrun, 1999, Mittler, 2002, Shao, et al., 2008a). Ground state triplet molecular oxygen (Fig. 8) may be converted to the much more reactive ROS forms either by energy transfer or by electron transfer reactions (Apel, Hirt, 2004). Transition metals, such as Fe and Cu, have catalytic function in these reactions (Florence 1984, Briat, Lebrun, 1999).

Under abiotic and biotic stresses, the increased generation of ROS initiates signaling responses that include enzyme activation, programmed cell death and cellular damage (Mittler, 2002, Neil et al., 2002, Pitzschke, Hirt, 2006). The production of ROS is connected not only with stress but also with common metabolic activities such as respiration and photosynthesis localized in mitochondria, chloroplasts, and peroxisomes (Fig. 9). NADPH oxidases, amine oxidases and cell-wall-bound peroxidases in the apoplast belong among other sources of ROS identified in plants (Mittler, 2002, Laloi et al., 2004). ROS serve also as signaling molecules, for example in the recognition of attack by fungal pathogens and herbivores (Bohnert, Sheveleva 1998, Mittler, 2002).

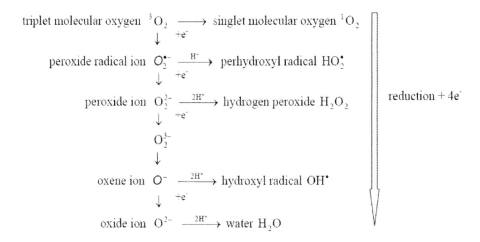

Figure 8. Generation of ROS by energy transfer or sequential univalent reduction of ground state triplet oxygen (adapted from Apel, Hirt, 2004).

Production and removal of ROS must be strictly controlled because they can cause damage to the biomolecules (e.g. membrane lipids, proteins, chloroplast pigments, enzymes, nucleic acids) and disturbances in signaling processes. In the case of stress induced by a variety of environmental stressors such as soil salinity, drought, extremes of temperature and heavy metals the equilibrium between ROS production and scavenging may be perturbed (Møller, 2001, Rodrígez-López et al., 2000, Mittler 2002, Rios-Gonzales et al., 2002, Zhu et al., 2004, Rhoads et al., 2006). The cellular redox perturbation seems to be an essential prerequisite for development of heavy metals - dependent phytotoxicity symptoms (Sharma, Dietz, 2009). The rapid increase in ROS concentration is called "oxidative burst" (Apel, Hirt, 2004).

To combat the oxidative damage plants have the antioxidant defense system comprising of enzymes catalase, peroxidases, superoxide dismutases and the nonenzymic constituents such as α-tocopherol, ascorbate and reduced glutathione which remove, neutralize and scavenge the ROS.

Heavy metals present in the environment can be divided into two groups: non-redox-active heavy metals, such as Zn and Cd and the redox-active heavy metals Fe, Cu, Cr, V and Co (Mithöfer et al., 2004). Redox-active heavy metals directly contribute to the formation of ROS. A general feature of heavy metals is to bind to proteins and other targets, or to compete for binding site in several functional groups (-SH means thiols, -COO$^-$ means carboxylic acids and imidazole means histidyl residues). This leads to the changes in functions of targeted proteins, which imply changes in cell metabolism, or trigger signaling events, which can lead to acclimation (Sharma, Dietz, 2009) (Fig. 10).

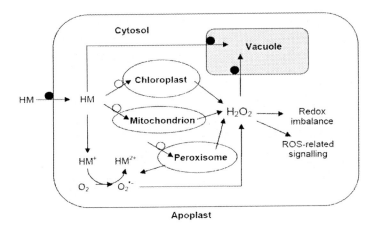

Figure 9. Pathways of heavy metal-dependent ROS generation (adapted from Sharma, Dietz, 2009) (HM – heavy metal ion).

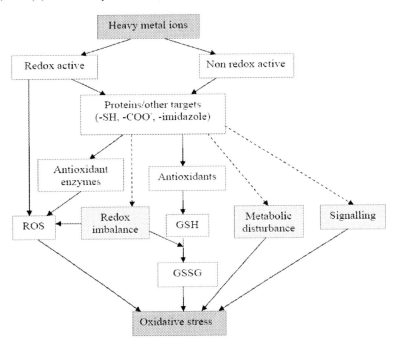

Figure 10. Mechanisms of heavy metal induced oxidative stress and related cellular processes (adapted from Rama Devi, Prasad, 2004, Sharma, Dietz, 2009).

Redox-active heavy metals cause the conversion of H_2O_2 to the highly reactive $OH^\bullet$ molecule via the Fenton reaction (1, 2). During this process metal ion is oxidized and consequently reduced with superoxide radicals $O_2^{\bullet-}$. Direct formation of $OH^\bullet$ from H_2O_2 and $O_2^{\bullet-}$ was suggested by Haber–Weiss (3) (Mithöfer et al., 2004, Hernández, et al., 2009, Sharma, Dietz 2009). Chemical reactions involved in hydroxyl radical $OH^\bullet$ generation are described as follows:

Fenton Reaction

$$Fe^{2+}/Cu^+ + H_2O_2 \rightarrow Fe^{3+}/Cu^{2+} + OH^\bullet + OH^- \tag{1}$$

$$Fe^{3+}/Cu^{2+} + O_2^{\bullet-} \rightarrow Fe^{2+}/Cu^+ + O_2 \tag{2}$$

Haber-Weiss Reaction

$$H_2O_2 + O_2^{\bullet-} \xrightarrow{\ Fe^{2+}, Fe^{3+}\ } OH^\bullet + OH^- + O_2 \tag{3}$$

ROS react with different cellular constituents. Their reactivity in cell depends on chemical reactivity, redox potential, half-life and mobility within the cellular compartments (Šlesak et al., 2007, Qiu et al., 2008).

The $OH^\bullet$ radicals are the most reactive and short-lived. This kind of ROS initiates radical chain reactions and probably causes irreversible chemical modifications of various cellular components, oxidation of biomolecules within diffusion distance, reversible as well as irreversible oxidative modifications of proteins (Taulavuori et al., 2005).

The hydroperoxyl radical ($HO_2^\bullet$), protonated form of superoxide radical $O_2^{\bullet-}$ is another ROS, that is probably responsible for the lipid peroxidation - an autocatalytic process that changes membrane structure and function resulting in an increase of the plasma membrane permeability which leads to leakage of potassium ions and other solutes and may finaly cause cell death (Chaoui et al., 1997, Verma, Dubey, 2003, Rai et al., 2004).

Likewise, DNA is oxidized mainly by $OH^\bullet$ and $^1O_2$, which have been reported to affect guanine, but less by H_2O_2 and $O_2^{\bullet-}$ and it also reacts with reactive nitrogen species. Oxidized polyunsaturated fatty acids are precursors for signaling molecules like jasmonic acid, oxilipins and volatile derivatives (Briat, Lebrun, 1999, Apel, Hirt, 2004, Mithöfer et al., 2004, Sharma, Dietz 2009).

ROS Detoxification – Antioxidant Systems

Various stressors destroy the equilibrium between ROS production and detoxification. To resist oxidative damage, the antioxidant enzymes and certain metabolites present in plants play an important role leading to adaptation and ultimate survival of plants during periods of stress (Pitzschke, Hirt, 2006, Shao et al., 2008b).

The role of antioxidant defence mechanisms is to interact with active forms of oxygen and to keep ROS at a low level and prevent them from exceeding toxic thresholds (enzymatic antioxidant systems) and regenerate oxidized antioxidants (non-enzymatic systems) (Larson, 1988, Apel, Hirt, 2004). Various antioxidant responses and degrees of tolerance to metal-induced oxidative stress are exhibited by different metal accumulating species (Qiu et al., 2008).

Non-Enzymatic ROS Scavenging Mechanisms

Non-enzymatic antioxidants include the major cellular redox buffers ascorbate, glutathione reductase (GR) and glutathione (GSH), as well as tocopherol, flavonoids, alkaloids and carotenoids (Foyer, Noctor 2005a, Shao et al., 2008a, Hernández et al., 2009, Sharma, Dietz 2009). GR and GSH are important components of the ascorbate-glutathione pathway responsible for the removal of H_2O_2 in different cellular compartments. Under stress-free conditions the glutathione pool is almost completely reduced. Under stress GSH is oxidized by ROS forming oxidized glutathione (GSSG), ascorbate is oxidized to monodehydroascorbate (MDA) and dehydroascorbate (DHA). Through the ascorbate-glutathione cycle GSSG, MDA and DHA can be reduced reforming GSH and ascorbate. In responses to the abiotic stress, plants increase the activity of GSH biosynthetic enzymes and GSH level (Zhang, Kirkham, 1996, Yadav, 2010, Potters et al., 2010). Glutathione has specific role within stress alleviation (Fig. 11). It acts in ROS scavenging as the part of ascorbate – glutatione cycle, secondly it is utilized by phytochelatin synthase in phytochelatin synthesis and finally even GSH molecule itself may conjugate directly with heavy metal ions and transport them into the vaculole (Rao, Sresty, 2000, Yadav, 2010).

The decrease of GSH/GSSG ratio reflect a precarious state of the cell in which sensitivity to stress is enhanced. It indicates a lower capacity of the cell to cope with ROS and then to overcome an oxidative stress (Karatas, et al., 2009).

Ascorbate (ASC) is one of the most important water-soluble antioxidants in plants. It is the most abundant contributor to general redox metabolism in plant cells. ASC biosynthesis is also directly tied into the redox network of a cell, and depends on the availability of oxidised cytochrome c from the mitochondrial respiration chain (Potters et al., 2010). Ascorbate along with glutathione (GSH) is involved in a fine regulation of the H_2O_2 level of the ascorbate-glutathione cycle. Ascorbate and glutathione are present in plant tissues in millimolar concentrations and in stress conditions their levels increase. Ascorbate reacts directly with ROS ($OH^\bullet$, $O_2^{\bullet-}$ and $^1O_2$), it also acts as a secondary antioxidant by reducing the oxidized form of α-tocopherol and preventing membrane damage (Noctor, Foyer, 1998, Apel, Hirt 2004).

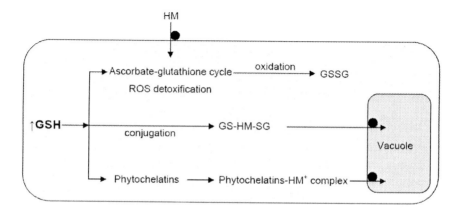

Figure 11. Glutathione-mediated detoxification of heavy metal stress response in plants (adapted from Anderson 1998, Foyer et al., 2001, Yadav 2010) (HM – heavy metal ion).

Enzymatic ROS Scavenging Mechanisms

Enzymatic ROS scavenging mechanisms in plants include ascorbate peroxidase (APX), glutathione peroxidase (GPX), superoxide dismutase (SOD) and catalase (CAT). These enzymes are involved in the detoxification of $O_2^{\bullet-}$ radicals and H_2O_2, thereby preventing the formation of $OH^\bullet$ radicals (Zhang, Kirkham, 1996).

SODs act as the first line of defense against ROS, dismutating superoxide to H_2O_2. CAT, APX and GPX subsequently detoxify H_2O_2. CAT converts hydrogen peroxide into water. In contrast to CAT, APX converts H_2O_2 into water by the ascorbate-glutathione cycle (an ascorbate is the reducing agent and glutathione (GSH) is the part of regeneration system). Similarly to APX, GPX also converts hydrogen peroxide into water by the glutathione-peroxidase cycle using equivalents from GSH as reducing agents. Oxidized GSSG is again converted into GSH by glutathione reductase (GR) (Tab. 3) (Apel, Hirt, 2004, Santos de Araujo et al., 2004). The ASC–GSH cycle is another important antioxidant mechanism involved in H_2O_2 detoxification that it is composed by the enzymes APX, GR, DHAR and MDHAR, and by GSH and ASC (Noctor, Foyer, 1998).

Table 3. The principal modes of enzymatic ROS scavenging by superoxide dismutase (SOD), catalase (CAT), ascorbat peroxidase, and the glutathione peroxidase (GPX) (adapted from Apel, Hirt 2004)

Superoxide Dismutase: $O_2^{\bullet-} \xrightarrow{SOD} H_2O_2$	(4)
Catalase: $H_2O_2 \xrightarrow{CAT} H_2O + 1/2 O_2$	(5)
Ascorbate peroxidase: $H_2O_2 + \text{ascorbate} \xrightarrow{APX} H_2O + MDA$	(6)
Glutathione peroxidase: $H_2O_2 + GSH \xrightarrow{GPX} H_2O + GSSG$	(7)

Among the number of enzymes regulating H_2O_2 intracellular levels CAT, APX and GPX are considered to be the most important (Noctor, Foyer, 1998, Milone et al., 2003).

CAT is less efficient than peroxidases (POD) in H_2O_2 scavenging because of its low substrate affinity. Therefore as long as the stress is not too strong for the plant's defense capacity, the main response to heavy metals is an increase in SOD and POD activities (Siedlecka, Krupa, 2002). Peroxidases are widely accepted as "stress enzymes". The measurement of their activity is one of the most common parameters in metal stress evaluation. Increasing of their activity was well documented under a variety of stressful condition, such as water stress (Yordanov, 2000), chilling, salinity (Rios-Gonzalez, 2002), γ-radiation (Verma, Dubey, 2003) and under toxic levels of metals (Baccouch et al., 1998, Shah et al., 2001). Furthermore, GPX participates in the lignin biosynthesis and might build up a physical barrier against poisoning of the heavy metals (Rai et al., 2004). Peroxidases are ubiquitous enzymes found in virtually all green plants, the majority of fungi and aerobic bacteria (Kvesitadze et al., 2006). The extent of oxidative stress in a cell is determined by the amounts of superoxide, H_2O_2 and hydroxyl radicals. So instead of only POD activity measurement SOD, APX and CAT activities measured together would provide more specific picture about the extent of oxidative stress at the cell level (Apel, Hirt, 2004).

3.5.3. Influence on Photosynthetic System

Photosynthesis, as the leading physiological process that determines primary production is often affected by heavy metal toxicity while roots are directly exposed to the contaminated environment (Gilbert et al., 2004, Durand et al., 2010). As photosynthesis is a very sensitive process to heavy metal contamination, plants respond quickly to prevent the photosynthetic system from suffering from irreversible damages (Zhou et al., 2009). Chloroplasts recognize the occurrence of abiotic stress and transfer stress signals to the rest of the cell (Foyer, Noctor, 2005b). The excess of heavy metals elicits firstly changes in the chloroplast ultrastructure and Fe deficiency (Slováková, Mistrík, 2007). Later the presence of heavy metals causes several negative effects on various photosynthetic processes, such as ·biosynthesis of chlorophyll, functioning of photochemical reactions, the activities of enzymes of the Calvin cycle and the photosynthetic CO_2 fixation (Hattab et al., 2009, Jia et al., 2010). Also ROS are produced in the chloroplast by the electron acceptor of photosystem I contributing to the high ROS level in the stressed cell (Hernández et al., 1995). During environmental stresses $NADP^+$ regeneration in the Calvin cycle is reduced followed by the over-reduction of the photosynthetic electron transport chain leading to the production of superoxide radicals and singlet oxygen in the chloroplasts (Shao et al., 2008b). It was reported that in the presence of heavy metals photosystem I is not, or only slightly, inhibited but the same concentration of heavy metals inhibits photosystem II completely (Atal et al., 1991). There is a general agreement that photoinhibition is primarily caused by an inactivation of the electron transport system in thylakoids predominantly affecting the reaction centers of PS II leading to a decreased photochemical efficiency (Vernay et al., 2007). It was found that the mechanism of the PS II inhibition might be a competitive inhibition of Ca^{2+} ions in catalytic center of PS II by divalent metal ions (Vrettos et al., 2001). Faller et al. (2005) studying mechanisms of Cd toxicity found that Cd^{2+} binding to the essential Ca^{2+} site during photoactivation is likely to be an important mode of action in vivo for inhibiting photosynthesis.

The effect of heavy metals on the efficiency of PS II probably depends on the plant genera and age, time of exposure to metals and metal content in leaf tissues (Ke, 2007).

The Mg substitution in the chlorophyll molecule by other metals is one of the key processes of heavy metal toxicity. It leads to inefficient light harvesting in the antenna complexes, inhibits electron transfer in the reaction centers, finally resulting in breakdown of photosynthesis (Prasad et al, 2001, Ventrella et al., 2009). Mg substitution rate represents more important factor in the decreasing of net photosynthesis than concentration of particular metal (Küpper et al., 2000). Küpper et al. (1996) found almost proportional relationship between toxicity (at constant concentration) and the tendency of an ion be bound in the center of the chlorophyll molecule. The substitution tendency decreased in the following order:

$$Hg^{2+} > Cu^{2+} > Cd^{2+} > Zn2^+ > Ni^{2+} > Pb^{2+}$$

Chlorophyll molecules with central metal ion such as Ag, Cu, Ni and Zn are stable but with Cd and Ni are unstable so plants bleach easily. Bleaching of plant leaves in the presence of metals is also connected with light intensity. In shady conditions the chlorophyll content of plants affected by the first group of metal forming stable substituted chlorophyll molecules (Ag, Cu, Ni, Zn) decreases only very slightly, plants color turns from green to blue-green. The color does not change, or change very little, but the photosynthesis is inhibited. Even dead plants may look vital but they are often fragile probably due to cellulose breakdown. Unstable substituted chlorophyll molecules (with Cd or Hg) are degraded so plants bleach in their presence. In contrast, by the intense light plant, in the presence of heavy metals, bleaches almost completely independently of the kind of metal present in their environment. Only a small part of chlorophyll (< 2%) was found to be accessible to substitution by heavy metal ions in sunlight (Küpper et al., 1996).

Metals such as Cu can cause the decrease of chlorophyll concentration also due to the peroxidation of chloroplast membrane (Farooqui et al., 1995, Prasad et al., 2001). Chloroplasts have a complex system of membranes rich in polyunsaturated fatty acids, which are potential targets for peroxidation (Hattab et al., 2009).

In normal conditions accessory pigments (carotenes and xanthophylls) function at normal conditions in pigmentation and as antioxidants under stress (Havaux, 1998, Škerget et al., 2005).

To evaluate the overall state of photosyntethic system the redox state of plastoquinone or chlorophyll *a* fluorescence can be measured (Potters et al., 2010). As the total chlorophyll content and relative content proportion of chlorophyll *a* and chlorophyll *b* are reduced through inhibition of chlorophyll biosynthesis, also these parameters can serve as the biomarkers of the heavy metal effects on plants. The measurement of chlorophyl *a* fluorescence can give a better picture about the chlorophyll as well as photosynthese status then only measurement of the chlorophyll content because heavy metal-substituted chlorophylls exhibit a blue-shift of a red absorbance maximum in the photosynthetic pigment extract (Küpper et al., 1996).

Chapter 4

STRESS EVALUATION

Evaluation of plant stress induced by heavy metals is an important parameter for the plant selection process in phytoremediation. Key parameters in plant stress evaluation are in greater detail described in following section.

4.1. GERMINATION

Germination assay is a basic procedure to determine heavy metals toxic effects on plants (An, 2006, Labra et al., 2006, Di Salvatore et al., 2008). It is well documented that germination process is highly disturbed by the metal stress; however, there are not much explanation on the molecular mechanism of the inhibition of seed germination in the presence of metals (Ahsan et al., 2007). Seed germination and the early seedling growth are more sensitive to metal pollution because some of the defense mechanisms have not developed (Liu et al., 2005, Xiong, Wang, 2005). Delay in germination has been often observed after heavy metals exposure (Espen et al., 1997, Bansal et al., 2002, El-Ghamery et al., 2003, Rodriguez, Alatossava, 2010).

In the study of Rahoui et al. (2010) the effect of cadmium on the germination of seeds and the growth of early seedlings of pea (*Pisum sativum*) was examined. Nutrient loss in germination medium resulting in the drastic decrease in their availability for an adequate metabolism acts, at least in part, to lower the germination rate and to delay the subsequent embryo growth in Cd-treated seeds. Results of Ahsan et al. (2007) obviously show that copper has a detrimental effect on rice seed germination and that 1.5 mM of copper is the end point for rice seed germination. The germination rate of the rice seeds significantly decreased with the increase of copper concentrations, ranging from 1.0 mM to 2.0 mM. The shoot growth of the germinating seeds was greatly inhibited at 0.2 mM copper, and further decreased at higher copper concentrations. In addition, no root formation was observed at copper concentrations higher than 0.5 mM. Relationship between the concentration of As in wheat seedlings and germination percentage, relative root length, relative shoot height exhibited significant negative correlation (Liu et al., 2005). Li et al. (2007) observed negative effects of arsenic on seed germination and the growth of roots and shoots only at higher arsenic concentrations (5-20 $mg.kg^{-1}$) but at low arsenic concentrations (0-1 $mg.kg^{-1}$) stimulation of seed germination was reported. Peralta et al. (2001) found that 40 $mg.kg^{-1}$ of Cr^{VI} reduced the ability of lucerne seeds (*Medicago sativa*) to germinate by 23% and decreased their growth in the contaminated medium. In according to Zeid (2001) the reduced germination of seeds under the Cr stress could be caused by a depressive effect of Cr on the activity of amylases and on the subsequent transport of sugars to the embryo axes. Sensitivity of the seed germination and root elongation test can be improved by replacing filter paper with agar (Di Salvatore et al., 2008).

Wierzbicka, Obidzińska (1998) found that the negative effect of metal on seed germination is strongly related to the seed coat permeability to metal ions. Ni, Cd and Al have the most negative effect on seed germination; they inhibit germination even in low concentrations. They are followed by Pb and Cu then less toxic Zn which in low concentrations supports seed germination of *Helianthus annuus* (Chakravarty, Srivastava, 1992). Combination of metals might have more detrimental effect on seed germination in comparison with single metals, e.g. germination rate of *Echinochloa colona* seeds declined when Cr + Ni were added together (Rout et al., 2000).

Adaptation of plants can be observed also at the seed germination level. For example, the seeds of *Echinochloa colona* collected from the contaminated area of minewaste dumps showed higher germination rate in metal containing solutions than the seeds of the same plant from uncontaminated areas. For these seeds higher germination was observed in solutions with metal in comparison with metal free solution (it was opposite to the seeds collected in uncontaminated area (Rout et al., 2000). The effect of selected heavy metals on seed germination is summarized in Tab. 4.

Table 4. Effects of heavy metal ions on seed germination

Metal form		Concentration*	Plant	Effect	Reference
Cu	Cu(NO₃)₂	0-1.024 mM	Lettuce, broccoli, tomato, radish	No effect, germination constant >90%	Di Salvatore et al., 2008
	CuSO₄	1.5 mM	Oryza sativa	Detrimental effect on germination	Ahsan et al., 2007
		1.0-2.0 mM		Decreased germination rate	
	CuCl₂	0.5-8.0 mM	Triticum aestivum Cucumis sativus	Decreased germination	Munzuroglu, Geckil, 2002
		8.0 mM		Germination rate reduced	
		1280 mg.kg⁻¹	Sorghum bicolor	Decreased germination	An et al., 2006
Zn	Zn(NO₃)₂	40 mg.kg⁻¹	Medicago sativa	Nonsignificantly reduced germination	Peralta et al., 2001
	ZnSO₄	0.031-0.155 mM	Nigella sativa, Triticum aestivum	Reduced germination	El-Ghamery et al., 2003
		0.155 mM		Completely inhibited germination	
		250, 500 mM	Helianthus annuus	Germination decreased by 80%, 88%	Chakravarty, Srivastava, 1992
		1000 mM		No germination	
Pb	PbCl₂	80-640 mg.kg⁻¹	Triticum aestivum Zea mays	Decreased germination	An et al., 2006
		0.5-8.0 mM	Triticum aestivum Cucumis sativus	Decreased germination	Munzuroglu, Geckil, 2002
	Pb(NO₃)₂	0-1.024 mM	Lettuce, broccoli tomato, radish	Seed germination constant >90%	Di Salvatore et al., 2008
		0.302,3.02 mM	Phaseolus vulgaris Pisum sativum	Delayed germination	Wierzbicka, Obidzinska 1998
				Completely inhibited germination	
		30.2 mM	Dianthuschinensis Brassica napus	Delay in germination, germination was not affected	
			Soja hispid, Cucumis sativus	Nonsignificant effect on germination	
Asᴵᴵᴵ	NaAsO₂	0.004, 0.031, 0.062 mM	Oryza sativa	Reduced germination by 8; 30; 83%	Abedin, Meharg, 2002
		0.05, 0.1 mM		Germination decreased by 9%, 59%	Shri et al., 2009

Table 4. (Continued).

Metal form		Concentration*	Plant	Effect	Reference
As^V	Na$_2$HAsO$_4$	0.123 mM	*Triticum aestivum*	Reduced germination by 38%	Liu et al.,2005
		0.0027, 0.022, 0.043 mM	*Oryza sativa*	Germination decreased by 3; 13; 61%	Abedin, Meharg, 2002
		0.086 mM	*Triticum aestivum*	Germination decreased by 24.2%	Liu et al., 2005
		0.1, 0.5 mM	*Oryza sativa*	Germination decreased by 6%, 53%	Shri et al., 2009
Co	CoCl$_2$	>7.5 mM	*Triticum aestivum*	Germination moderately decreased	Munzuroglu, Geckil, 2002
			Cucumis sativus	Germination not affected	
		50 mM	*Phaseolus vulgaris*	Reduced germination by 67%	Zeid, 2001
		100 mM		Completely inhibited germination	
Cr^{III}	CrCl$_3$	50 mM	*Phaseolus vulgaris*	Reduced germination by 34%	Zeid, 2001
Cr^{VI}	K$_2$Cr$_2$O$_7$	0.004, 0.008 mM	*Echinochloa colona*	Decrease of germination (96%, 78%)	Rout et al., 2000
		40 mg.kg^{-1}	*Medicago sativa*	Reduced germination by 50%	Peralta et al., 2001
Ni	Ni(NO$_3$)$_2$	0-1.024 mM	Lettuce, broccoli tomato, radish	Seed germination constant >90%	Di Salvatore et al., 2008
	NiSO$_4$	0.0081, 0.016 mM	*Echinochloa colona*	Decrease of germination (92%, 76%)	Rout et al., 2000
		<0.08 mM	*Raphanus sativus*	No effect on germination	Espen et al., 1997
		0.1-0.2 mM		Inhibition of germination	
		0.4 mM		Completely inhibited germination	
Cd	Cd(NO$_3$)$_2$	0-1.024 mM	Lettuce, broccoli tomato, radish	Seed germination constant >90%	Di Salvatore et al., 2008
	CdCl$_2$	0.5-8.0 mM	*Triticum aestivum, Cucumis sativus*	Decreased germination	Munzuroglu, Geckil, 2002
		5 mM	*Pisum sativum*	Lower germination rate, delayed embryo growth	Rahoui et al., 2010

* Metal concentrations in solutions are expressed in units mol.dm^{-3} and in soils or substrates in mg.kg^{-1}.

The effect on germination is strongly affected by the metal form. In general, metals added into soil or solution as nitrates had no or non-significant effect on seed germination irrespectively on kind of metal in the compound. Even Cd added as nitrates had no effect on germination. Addition of the same metal but in the form of chlorides or sulfates had detrimental effects on germination. Metals, which in aqueous solution form oxoanions (As, Cr), have always decreased the seed germination.

4.2. REDUCTION IN GROWTH

Toxic metals are usually responsible for the reduction of plant growth (Chaoui et al., 1997). The reduction in growth can be expressed as reduced growth rate or decreased biomass production. This reduction can be due to specific toxicity of the metal to the plant, antagonism with other nutrients in the plant, or inhibition of the root penetration in the soil (Begonia et al, 1998).

To test the phytotoxicity of metals Leita et al. (1993) suggested calculation of the Grade of Growth Inhibition (GGI).

$$GGI = [(C-T) / C],$$

where C and T represent the dry weight of tissues of control (C) and metal-treated plants (T).

For plants without stress, where growth is not inhibited the GGI = 0, i.e., 100% growth.

The decrease of dry biomass weight was reported for cadmium treated pepper (León et al., 2002), bean (Chaoui et al., 1997) and wheat (Milone et al., 2003). But there were reported also cases when a low level of cadmium had positive effect on plants growth although they are very poorly discussed in literature (Arduini et al., 2004). Two possible mechanisms were suggested. Low cadmium levels hyperpolarize the plasma membranes at the root surface, thus increasing the trans-membrane potential, which is an energy source for cation uptake. Moreover, cadmium has been found to induce genes related to mammalian cell proliferation, which could increase growth, though they are also considered responsible for Cd-induced carcinogenesis (Arduini et al., 2004). Marchiol et al. (2004) did not observed any obvious symptoms of metal toxicity during the treatment with the mixture of Cd, Pb, Cr, Ni, Zn and Cu but they have recorded growth reduction and decrease in biomass production in both plants, *Brassica napus* and *Raphanus sativus,* treated with the metal mixture.

Also Begonia et al. (1998) for Indian mustard (*Brassica juncea*) observed no growth and biomass inhibition but even higher amount of biomass in plants treated with lead. The Indian mustard plants treated with lead in concentration 100 and 250 mg/l produced 20% more biomass than control plants. And at very high lead concentration in solution (500 mg/l) the biomass production decreased very little, only by 7% in comparison with control. It is possible that the small amount of Pb translocated to the shoot was not enough to elicit reduction in shoot biomass. Sekhar et al. (2004) recorded only a slight decrease in the growth of *Hemidesmus indicus* root and shoots treated with Pb concentration from 100 to 10 000 mg.kg^{-1}.

For *Pteris vittata* treated with As and Cd and As and Pb stimulated growth and 12 times higher biomass production in comparison with controls treated only with arsenic was observed (Fayiga et al., 2004). The authors suggested that growth stimulation might result from added N nutrition since all metals were added as nitrate salts. Kadukova et al. (2008) observed stimulation of biomass production in Pb, Cd and Pb + Cd treated *Tamarix smyrnensis.* The authors suggested that growth stimulation by Cd and Pb might also result from added N nutrition. On the basis of published experimental results it is obvious that the plant growth does not need to be always reduced in the presence of heavy metals so the determination of only growth rate or biomass production can not be considered as the only one parameter to evaluate the plant stress. The growth reduction and decrease in biomass production after selected metal addition into growth media (soil/hydroponic media) are listed in Tab. 5.

Table 5. The growth rate reduction and the biomass production decrease in the presence of metals

Metal form		Concentration*	Plant	Effects	Reference
Pb	$PbCl_2$	80-640 mg.kg⁻¹	*Triticum aestivum* *Zea mays*	Plant growth sensitive to the toxicity endpoint	An et al., 2006
		50,200 mg.kg⁻¹	*Pteris vittata*	Stimulation of biomass production	Fayiga et al., 2004
	$Pb(NO_3)_2$	800 mg.kg⁻¹	*Tamarix smyrnensis*	Stimulation of biomass production	Kadukova et al., 2008
		100-10000 mg.kg⁻¹		Root and shoot growth only slightly affected	Sekhar et al., 2004
		3.02×10^{-4} mM	*Brassica juncea*	Roots of plants were not affected	Begonia et al., 1998
		7.55×10^{-4} mM		Reduction of total leaf area by 25% and root growth by 21%	
		0.024-0.048 mM	*Arabis paniculata*	Slight stimulation of the shoot biomass production	Tang et al., 2009
	$CdSO_4$	0.024-0.048 mM	*Triticum durum*	Root growth inhibition	Milone et al , 2003
	$CdCl_2$	0.01-0.05 mM	*Pisum sativum*	Significant growth inhibition	Sandalio et al., 2001
		0.05 mM		Decreased dry weight of leaves	
		0.125-0.5 mM	*Capsicum annuum*	Significant reduction in the growth, biomass production	León et al., 2002
Cd	$Cd(NO_3)_2$	16 mg.kg⁻¹	*Tamarix smyrnensis*	Stimulation of biomass production	Kadukova et al., 2008
		50-200 mg.kg⁻¹	*Pteris vittata*	Significantly increased biomass production	Fayiga et al., 2004
		0.0011, 0.0021 mM	*Miscanthus sinensis*	Increased root length	Arduini et al., 2004
		0.0032 mM		Root length markedly reduced, root biomass no significantly affected	

Table 5. (Continued).

Metal form		Concentration*	Plant	Effects	Reference
		0.0011-0.0032 mM	Vetiveria zizanioides	No effect on biomass production first 30 days, after that reduction by 36%	
	CdCO$_3$	0.0043 mM	Vetiveria zizanioides	Decrease of shoot dry weight (17.3%) and root dry weight (32.5%)	Xu et al., 2009
	CdCl$_2$	0.009-0.089 mM	Arabis paniculata	Significant stimulation of the shoot biomass production and root growth	Tang et al., 2009
		0.031-0.155 mM	Nigella sativa, Triticum aestivum	Inhibited root growth	El-Ghamery et al. 2003
		0.1 mM	Phaseolus vulgaris	Significant reduction of growth in roots (40%) and in stems (28%)	Chaoui et al., 1997
Zn	ZnSO$_4$	0.306-0.612 mM	Arabis paniculata	Statistically significant increase in shoot dry weight	Tang et al., 2009
		0.612-2.447 mM	Arabis paniculata	Slight stimulation of the shoot biomass production	
		10 mM	Helianthus annuus	Supporting 100% growth	Chakravarty, Srivastava, 1992
	ZnCO$_3$	80 mg.kg^{-1}	Vetiveria zizanioides	Decrease of shoot dry weight (14.2%) and root dry weight (14.1%)	Xu et al., 2009
		0.0002 mM		15% increase in plant biomass	
		0.002-0.02 mM	Brassica juncea	Significant reduction in plant biomass production	Pandey et al., 2005
Cr^{6+}	K$_2$Cr$_2$O$_7$	0.0068-0.034 mM	Salvinia minima	Dry plant biomass weight significantly reduced	Prado et al., 2010
		0.05 mM	Datura innoxia	Reduction of biomass production	Vernay et al., 2008
		10 mg.kg^{-1}	Medicago sativa	Increase root size by 37%	Peralta et al., 2001
		40 mg.kg^{-1}		Decreased growth	

Metal form		Concentration*	Plant	Effects	Reference
Cu²⁺	CuSO₄	0.01, 0.05 mM	Prunus cerasifera	Relative growth rate not modified	Lombardi et al., 2005
		0.1mM		Significant growth reduction	
		0.2 mM	Oryza sativa	Shoot growth inhibited	MunzurogluG cckil,2002
		≥0.5 mM		No root formation	
	CuCl₂	0.001, 0.0012 mM	Urochloa mosambicensis	50% decrease of roots and shoots	Kopittke et al. 2009
		0.01 mM	Hordeum vulgare	Shoot height, root length, dry weight of shoots and roots reduced	Guo et al., 2007
	Cu(NO₃)₂	0-1.024 mM	Lettuce, tomato broccoli, radish	Root elongation	Di Salvatore et al., 2008
Ni²⁺	NiCl₂	0.06, 0.12 mM	Matricaria chamomilla	No significant effect on growth	Kováčik et al., 2009
	NiSO₄	0.017 mM	Nasturtium officinale	Increased biomass production	Duman, Ozturk,2010
		0.426 mM		Significant decrease in biomass production	
	NiCl₂	0-168 mg.kg⁻¹	Lactuca sativa	Biomass production not affected negatively	Poulik, 1999
		≥28 mg.kg⁻¹	Solanum lycopersicum	Dry matter weight decreased	
		≥84 mg.kg⁻¹		Significant increase of biomass production	
		50-3200 mg.kg⁻¹	Lycopersicon esculentum Hordeum vulgare	Barley root length and tomato shoot biomass significantly (>70%) reduced	Rooney et al., 2007
Co²⁺	CoSO₄	0.05 mM	Phaseolus aureus	Enhancement in growth, increase of biomass production	Tewari et al., 2002
		>0.1 mM		Decreased growth and dry matter yield	
	CoCl₂	≥7.5mM	Triticum aestivum, Cucumis sativus	Completely inhibition of root growth	Munzuroglu Gcckil,2002
		160-1600 mg.kg⁻¹	Lycopersicon esculentum Hordeum vulgare	Plant shoot biomass production significantly inhibited	Li et al., 2009

Table 5. (Continued).

Metal form	Concentration*	Plant	Effects	Reference
As^{5+}	0.022-0.086 mM	Triticum aestivum	Reduced root length and shoot height	Liu et al., 2005
	>0.5 mM	Oryza sativa	No root formation	Shri et al., 2009
	0.5 mM		Root and shoot length, biomass greatly inhibited	
	5-100 mg.kg^{-1}	Pteris vittata	Growth and shoot biomass significantly enhanced	Tu, Ma 2005
	≥300 mg.kg^{-1}		Decreased biomass production	
	100 mg.kg^{-1}	Scutellaria baicalensis	Stimulated growth of roots, no significant effect on growth of shoots, increased total biomass	Cao et al., 2009
	200 mg.kg^{-1} >400 mg.kg^{-1}		Decreased plant biomass production	
As^{3+}	0.031-0.123 mM	Triticum aestivum	Reduced root length and shoot height	Liu et al., 2005
	0.05 mM	Oryza sativa	Root and shoot length, biomass greatly inhibited	Shri et al., 2009
	0.1 mM		Root weight and root length negatively affected	
	<0.001 mM	Nasturtium officinale	Increase in relative growth rate, stimulated plant growth	Ozturk et al., 2010
	0.003-0.01 mM		Negative effect on plant growth	
	0.05 mM		Significantly decrease of plant growth	
	0.015-0.077 mM	Lycopersicum esculentum Phaseolus vulgaris	Reduced root and shoot biomass production	Carbonell-Barrachina et al., 1997

Metal form		Concentration*	Plant	Effects	Reference
Cr	KCr(SO$_4$)$_2$	0.2 mM	*Datura innoxia*	Decrease of shoot, root biomass production	Vernay et al., 2008
	Cr(NO$_3$)$_3$	0.21-0.84 mM	*Miscanthus sinensis*	Biomass production decreased, stimulated root elongation	Arduini et al., 2006
		$\geq$0.63 mM		Growth completely stopped	

*Metal concentrations in solutions are expressed in units mol.dm^{-3} and in soils or substrates in mg.kg^{-1} .

The metal form (especially anion) does not always have so significant influence on the plant growth as it has on germination. Although in some cases when metal was added as nitrate higher biomass production and growth rate was observed in comparison with the decrease of biomass production and growth rate when the same metal was added into substrate as sulfate or chloride. But at high metal concentration the negative effect of metal cation usually predominates over the influence of anion. The influence of metal on plant growth and biomass production significantly depends on the plant species (if it is tolerant or not) and metal concentration.

4.3. PHOTOSYNTHETIC PIGMENTS

Chlorophylls are the basic photosynthetic pigments. They are essential for radiance fixation and consequent transformation of energy in photosynthesis. Carotenoids serve as accessorial pigments. They include carotenes and xanthophylls. They function both in pigmentation and as antioxidants (Havaux, 1998, Škerget et al., 2005). Carotenoids act as light-harvesting pigments and protect chlorophyll and membrane destruction by quenching triplet chlorophyll and removing oxygen from the excited chlorophyll-oxygen complex (Ekmekçi et al., 2008). The content and composition of photosynthetic pigments are important indicators of the status of photosynthetic apparatus and depend on the plant species, mineral nutrition and growth conditions (Masarovičová, Repčák, et al., 2008).

A common response of plants to the metal stress is a decrease of the chlorophyll content in leaves of plants and subsequently the reduction of photosynthesis that finally leads to a lower biomass production (Monteiro et al., 2009). The reduction of chlorophyll content in the presence of heavy metals may be due to an inhibition of chlorophyll biosynthesis (Xiong, 1997). The reduction of in the levels of photosynthetic pigments, including chlorophylls a and b and carotenoids, on exposure to heavy metals have been observed in many species for Cu (MacFarlane, Burchett, 2001, Prasad et al., 2001), Zn (Ghnaya et al., 2009, Radić et al., 2010), Cd (Wu et al., 2003, Ekmekçi et al. 2008), Pb (Mishra et al., 2006, Cenkci et al., 2010). Ghnaya et al. (2009) found the decrease of chlorophyll a, chlorophyll b contents and the total amount of chlorophyll in Brassica napus treated with different Cd and Zn concentrations. Likewise, Monferrán et al. (2009) observed symptoms of changes in the photosynthetic apparatus in Potamogeton pusillus after the exposure to copper, except of other changes he observed a decrease in chlorophyll *a* and chlorophyll *b*, Zaman and Zereen (1998) observed a significant decrease in chlorophyll *a*, chlorophyll *b* contents and the total amount of chlorophyll in radish plants treated with different Pb and Cd concentrations. The reduction in chlorophyll content in turn, at least partly, would lead to a decrease of biomass (Sinha et al., 1993). In shoot tissues of *Bacopa monnieri* a negative significant correlation of Fe uptake with chlorophyll content was noticed. Adverse effect of Fe on the chlorophyll contents might be due to a strong oxidation of the photochemical apparatus, reduction in chloroplast density and size, phosphorus deficiency or reduced Mn transport (Sinha et al., 2009). The exposure of pea *Pisum sativum* seedlings to Cd resulted in the reduction of chlorophyll and carotene contents in leaves. The deleterious effect of Cd became more pronounced with increasing concentrations. In plants exposed to 7 mg Cd/kg growth media, chlorophyll *a*, chlorophyll *b* and carotene decreased by 50.6%, 51.9% and 45.3%, respectively, compared to control plants (Hattab et al., 2009). Similarly, in green algae *Scenedesmus quadricauda* a slight decrease of chlorophyll content was observed after exposition to Cu (Kováčik et al., 2010).

In higher plants there are two kinds of chlorophyll – chlorophyll *a* and chlorophyll *b*. Their contents in plants are usually in the ratio 3:1 (chlorophyll *a* dominates). The changes in the ratio of chlorophyll *a/b* can also reflect the negative influence of heavy metals to the photosynthetic apparatus of plants (Porra, 2002). It can be connected with the fact that chlorophyll *a* is more sensitive to some heavy metals than chlorophyll *b* (Pandey, Sharma, 2002). A significant increase was seen in the chlorophyll *a/b* ratio for Zn alone, suggesting that the chlorophyll *b* pool is more sensitive to Zn exposure (Macfarlane, Burchett 2001). Dazy et al. (2008) found that Cr exposure disturbed the cellular redox status, chlorophyll content and chlorophyll *a/b* ratio in *Fontinalis antipyretica* apices. The effects on chlorophyll contents and chlorophyll *a/b* ratios were also shown even at low Cr concentrations.

Similarly to the case of growth reduction, heavy metals do not always cause the reduction in chlorophyll content. For example, Xiong (1997) did not observe any significant decrease of the chlorophyll content in plants *Sonchus oleraceus* treated with 800 and 1600 mg.kg^{-1} of lead. Only at a very high Pb concentration in soil (3200 mg.kg^{-1}) 18% of the decrease of chlorophyll content was observed. Also Gupta et al. (2009) did not report any changes in the chlorophyll content in the leaves of *Sesamus indicum* treated with the mixture of cadmium, copper and zinc. The chlorophyll *a*, chlorophyll *b* and carotene contents in seedling of *Pisum sativum* exposed to copper decreased only after addition of the highest Cu concentration -700 mg kg^{-1} (Hattab et al.,

2009). Similarly, only higher concentration of arsenic in soil (10 and 50 mg/kg) led to the decrease of chlorophyll *a*, chlorophyll *b* and carotenoids contents in clover shoots. After addition of zinc and cadmium together with arsenate in the heavy metal mixture, the variant chlorophyll content did not decreased so significantly as if only arsenate was added (Mascher et al., 2002). Exposition of *Sedum alfredii* leaves to 1000 μM of Cd in the nutrient solution resulted in the increase of the total chlorophyll, chlorophyll *a* and *b* contents on the basis of fresh weight, but the ratio of chlorophyll *a* to chlorophyll *b* decreased (Zhou, Qiu, 2007). Dazy at al. (2008) found that two different forms of chromium had the opposite effect on the chlorophyll content of *Fontinalis antipyretica*. Cr^{III} as nitrate salt seemed to decrease the total chlorophyll content whereas Cr^{III} as chloride salt and Cr^{VI} lead to chlorophyll accumulation. Cr^{III} modified chlorophyll *a/b* ratio but Cr^{VI} had no effect on that parameter.

Jiang et al. (2007) observed increasing of the chlorophyll content after applying external P in the complex pollution of Cd and Zn.

The decrease in carotenoid contents was found in rapeseed cultivar (*Brassica napus*) treated with Zn^{2+} and Cd^{2+}. In accordance to carotenoid content measurement these cultivars were more sensitive to the stress of Cd^{2+} then Zn^{2+}. This drastic reduction of carotenoids may be the result of a strong production of ROS (Ghnaya et al., 2009). In the study of Singh et al. (2008) carotenoid contents in *Beta vulgaris* plants decreased significantly with increasing concentrations of fly ash as compared to the control at 35 days after sowing. Increasing concentration of Cu significantly inhibited carotenoid concentration in *Withania somnifera* (Khatun et al., 2008).

In the case that metal were added into soil the carotenoids decrease was observed. But in plants growing in polluted areas a higher carotenoid content was found. It can be connected with the fact that carotenoid serve as antioxidants which could protect these plants against harmful effects and be part of adaptation mechanisms (Sillanpää et al., 2008). The influence of heavy metals on photosynthetic pigments is listed in Tab. 6.

Table 6. The influence of metals on photosynthetic pigments

Metal form		Concentration*	Effects	Plant	Reference
Pb	CH₃COOPb	<1600 mg.kg⁻¹	No significant effect	Sonchus oleraceus	Xiong 1997
		3200 mg.kg⁻¹	Significant decrease of chlorophyll by 18%		
	Pb(NO₃)₂	0.5-5 mM	Significant decrease of chl a, chl b, and carotenoids	Brassica rapa	Cenkci et al., 2010
		5 mM	Maximum inhibition of chl a (57.4%), of chl b (68.5%) and of carotenoids (38.1%)		
Cu	CuSO₄	0.001 mM	Stimulated chl a and chl b fluorescence	Lemna trisulca	Prasad et al., 2001
		0.002-0.05 mM	Fluorescence of chl a and chl b dramatically decreased		
		0.025, 0.05 mM	Decrease of chl a and chl b content and carotenoids		
		>1 mM	Marked decline of chl b fluorescence		
		0.5 mM	Carotenoid content decreased 71.4%	Zea mays	Tanyolaç et al., 2007
		0.5 - 1.5 mM	Significantly decreased chl $(a + b)$ and carotenoids contents		
		1.5 mM	Total chlorophyll content decreased about 72%		
	CuCl₂	≥0.149 mM	The highest drop in chl a and chl b (7 d)	Potamogeton pusillus	Monferrán et al., 2009
		≥0.297 mM	Significantly increased chl b (1 d); fall chl a, chl b (3 d)		
		0.743 mM	Increased chl a (1 d)		
		>1.487 mM	Significant declines in chl a and chl b and total chlorophyll	Avicennia marina	MacFarlane, Burchett, 2001
		5.948 mM	Carotenoids significantly reduced		

Table 6. (Continued).

Metal form		Concentration*	Plant	Effects	Reference
As[III]	NaAsO$_2$	0.001-0.05 mM		Increased carotenoids	Ozturk et al., 2010
		≤0.003 mM	Nasturtium officinale	Increased slightly photosynthetic pigments	
		0.003 mM		Increased carotenoids by 46%	
		0.005-0.05 mM		Decreased chl a and chl b	
As[V]	Na$_2$HAsO$_4$	0.133 mM	Pteris vittata Pteris ensiformis	No significantly reduced chll a, b and total chlorophyll	Rahman et al., 2007
		0.267 mM		Significantly reduced total chlorophyll	
		≤10 mg.kg^{-1}	Oryza sativa	No significantly affected chlorophyll content	
		10-30 mg.kg^{-1}		Decreased chl a and chl b	
		5 mg.kg^{-1}	Trifolium pratense	No changes in chlorophyll and carotenoids contents	
		10, 50 mg.kg^{-1}		Significantly reduced chlorophyll by 14% and carotenoid contents by 12%	
Zn	ZnSO$_4$	0.15, 0.3 mM	Lemna minor	Significant decreases of chl a, chl b and carotenoid contents	Radić et al., 2010
		2 mM	Brassica napus	Decreased chl a, chl b and carotenoid content	Ghnaya et al., (2009)
	ZnCl$_2$	0.01-0.2 mM	Ceratophyllum demersum	Maintained levels of chlorophyll and carotenoids in plant	Aravind, Prasad, 2004
		0-3.666 mM	Avicennia marina	Significant decline of chl a, chl b and total chlorophyll	MacFarlane, Burchett, 2001
		>3.666 mM		Significantly increased chlorophyll a/b ratio	
		7.331 mM		Only carotenoids significantly reduced	

Metal form		Concentration*	Plant	Effects	Reference
Cr^III	KCr(SO₄)₂	0.05-2 mM	*Datura innoxia*	Slight decrease of chlorophyll contents	Vernay et al., 2008
		2 mM		Decrease of chl *a*, chl *b* and total chlorophyll	
	CrCl₃	6.25×10^{-4} – 0.625 mM	*Fontinalis antipyretica*	Decrease of chlorophyll *a+b*	Dazy et al., 2008
		6.25-50 mM		Increased chlorophyll level	
	Cr(NO₃)₃	6.25×10^{-5} – 12.5 mM	*Fontinalis antipyretica*	Decrease of chlorophyll *a+b*	
		25, 50 mM		Increase of chlorophyll *a+b*	
		$\leq 6.25 \times 10^{-4}$ mM	*Fontinalis antipyretica*	Increased chlorophyll contents	Dazy et al., 2008
		$\geq 6.25 \times 10^{-3}$ mM		No changes in chlorophyll *a+b*	
		0.0002 mM		Increased chlorophyll (5 d)	Pandey et al., 2005
		0.002 mM		Increased carotenoids (15 d)	
Cr^VI	K₂Cr₂O₇	0.02 mM	*Brassica juncea*	Significantly increase of chlorophyll and carotenoids	
		0.01-0.1 mM	*Ocimum tenuiflorum*	Significantly reduced total chlorophyll, chl *a*, chl *b* and carotenoids	Rai et al., 2004
		0.1 mM		Reduced total chlorophyll, chl *a*, chl *b* and carotenoids	
		0.017-0.68 mM	*Pistia stratiotes Glycine max*	Decrease of chlorophyll and carotenoids	Ganesh et al., 2008
Ni	NiSO₄	0.001 mM	*Elodea canadensis*	Slight increase (15–20%) of pigment concentrations	Maleva et al., 2009
		<0.01 mM		No significant changes in chlorophyll and carotenoids	
		0.05 mM		Decreased chl *a* content to 75% Chl *a/b* ratio decreased by 15%	

Table 6. (Continued).

Metal form		Concentration*	Effects	Plant	Reference
	NiCl$_2$	0.0001-0.5 mM	Increasing chl a and chl b Carotenoids changed slightly Ratio chl a/b relatively constant	Spirodela polyrhiza	Appenroth et al., 2010
		≥0.1 mM	Significantly decreased ratio chl a/b	Lemna minor	
		0.1 mM	Decreased carotenoids to 86% (8 d) and 76% (13 d)	Zea mays	Drążkiewicz Baszyński, 2010
		0.2 mM	Decreased carotenoids to 78% (8 d) and 68% (13 d)		
		0.2 mM	Reduced pigments to 55% (8 d) and 54% (13 d)		
Co	CoSO$_4$	0.0001-0.05 mM	Increased chlorophylls and total carotenoids	Phaseolus aureus	Tewari et al., 2002
		0.1-0.4 mM	Decreased photosynthetic pigments		
		0.5 mM	Significant decrease in chll $a+b$ Decrease of ratio chl a/b	Brassica oleracea	Pandey, Sharma, 2002
Cd	CdCl$_2$	0.0001 mM	No significant increase in chlorophyll content	Hordeum vulgare	Wu et al., 2003
		0.001 mM	Slight decrease in chlorophyll		
		0.005 mM	Marked decline in chlorophyll		
	Cd(NO$_3$)$_2$	0.1 mM	Reduction of chl a by 41% and chl b by 43% in expanded leaves	Lactuca sativa	Monteiro et al., 2009
		0.3-0.9 mM	Significant reduction of chl a, chl b and chl ($a+b$) content	Zea mays	Ekmekçi et al., 2008

*Metal concentrations in solutions are expressed in units mol.dm^{-3} and in soils or substrates in mg.kg^{-1} .

The changes in the photosynthetic pigments levels depend on both metal concentration and time of exposition. At lower metal concentration no changes or even the increase of the total chlorophyll, chlorophyll *a*, chlorophyll *b* and carotenoids contents was often observed. In some case although the increase of the chlorophylls contents were observed at the beginning, their decrease after longer exposition period followed. Carotenoids often respond to the presence of metals later than chlorophylls suggesting that chlorophylls should be more sensitive to the heavy metal presence. Changes in the chlorophyll *a*/chlorophyll *b* ratio depend on the higher sensitivity of particular pigment (chlorophyll *a* or chlorophyll *b*) to metal added so they are different for different metals.

4.4. ANTIOXIDANT ENZYMES

Induction in the activities of antioxidant enzymes is a general strategy adopted by plants to overcome the oxidative stress due to the imposition of environmental stresses (Shah et al., 2001). Mechanisms of decreasing of the oxidative stress in plants by antioxidant enzymes are described above in the section 3.5.2. Measurement of the activity of antioxidant enzymes can provide useful information about the stress level in plants. The level of anionic isoenzymes of guaiacol dependent peroxidase represents both quantitative and qualitative changes caused by the metals and analyses of their profile can be considered a metabolic indicator of the heavy metal stress (Baccouch, 1998, Verma, Dubey, 2003, Qiu, et al., 2008). Significant increases in peroxidase activity were found with Cu and Zn at concentrations lower than those inducing visible toxicity (Macfarlane, Burchett, 2001). Cd has been found to trigger the oxidative stress in tobacco cells within minutes of exposure, possibly signaling the first defenses against heavy metals the plasma membrane (Boominathan, Doran, 2003). Ni induced high POD activity in leaves of *Zea mays* (Baccouch et al., 1998). Verma and Dubey (2003) found 1.2 – 5.6 times increase in peroxidase activity in roots of rice treated with lead. In the study of Kavuličova et al. (2009) guaiacol peroxidase activity strongly responsed to the metal exposure. Specific activity of peroxidase was found to be significantly higher (1.5 times) at plants treated even with low content of Cu, Zn and Cd in soil in comparison with control plants. At high metal content (5 times higher metal concentration than the standard in soil) guaiacol POD activity was 3.5 times higher than in control. The results indicate a considerable enhancement in the activity of guaiacol peroxidase, suggesting that this antioxidant enzyme acts to reduce the impact of metal toxicity.

In some plants the activity of peroxidase differs in shoots and in roots. For example, in tomato after Cu treatment high peroxidase activity was recorded only in roots and stems but not in shoots (Mazhoudi et al., 1997). Shah et al. (2001) observed the increase of guaiacol peroxidase activity in roots and shoots of rice treated with Cd. They recorded higher POD activity in shoots than in roots. On the contrary, Chaoui et al. (1997) did not observed the guaiacol- dependent peroxidase activity increase in the leaves of beans after Cd treatment.

Israr and Sahi (2006) investigated the effect of mercury on the levels of antioxidant enzymes SOD, APX and GR in *Sesbania drummondi* cell cultures. *Sesbania* cell cultures tolerated Hg up to a content of 40 µM, but higher contents caused toxicity to the cells. The activities of APX, SOD were markedly increased in response to Hg treatments. The activity of SOD in plants *Ceratophyllum demersum* increased 3.6-times in plants supplemented with Zn (200 µM) over the control but just 1.5-times in plants treated with Cd (10 µM) (Aravind, Prasad, 2003). In *Bruguiera gymnorrhiza* plants under the heavy metal stress (different concentration of Pb^{2+}, Cd^{2+}, and Hg^{2+}), CAT activity was not affected in the leaves, but increased in the roots. In *Kandelia candel*, CAT activity increased in both leaves and roots (Zhang et al., 2007).

At low metal concentration, when no toxicity symptoms were observed, no changes in antioxidant enzyme activity were observed for Mn treated barley plants and a slight but significant increase of SOD and membrane-bound APX activities and the decrease of soluble APX and CAT activities for plants treated with Cu were observed. But when barley plants were exposed to Mn and Cu concentrations which caused visible toxicity symptoms, in both cases the decrease of SOD and the increase of CAT and GPX activities were reported. The activity of APX after treatment with Cu did not change and after Mn treatment it decreased (Demirevska-Kepova, et al., 2004). In cyanobacteria *Spirulina platensis* the increase of SOD activity was observed when different heavy metals (Cu, Zn, Pb) were added into solution. The highest increase was observed when Pb was added (Choudhary et al., 2007). In the presence of Pb CAT activity in *Sesbania* plants was elevated by 265% and SOD activity by 180%, but no significant changes in their activities were observed in the presence of Pb + EDTA, Pb + DTPA or Pb + HEDTA (Ruley et al., 2004). Cadmium induced the oxidative stress in pea plants characterized by reduction in reduced GSH content, CAT (close to 50%) and CuZn-SOD activities. The effect of cadmium on APX was different from that of CAT, showing a slight increase in activity (Romero-Puertas et al., 2007). Cd treated pepper expressed the decrease of CAT and SOD and the increase of GPX (León et al., 2002). Zhang et al. (2005a) observed the decrease of CAT and SOD activities in garlic treated with 5 and 10 mM $CdCl_2$ at the beginning of the experiment but the activity of CAT slowly recovered to the extent of control or even increased over the control. When exposed to 1mM $CdCl_2$ CAT and SOD activities increased. POD activities at 5 and 10 mM Cd firstly dramatically increased and then slowly decreased but still were higher than in controls, at 1 mM POD activities gradually increased. High Cu concentration (2.5 mM) in media increased SOD activity in red cabbage seedlings but decreased CAT activity. On the contrary, lower Cu concentration slightly but statistically significantly increased CAT activity and did not change SOD activity (Posmyk, et al., 2009). Suppression of the activities of SOD, POD and CAT led to H_2O_2 burst, lipid peroxidation, cell death and growth inhibition in Cd treated rice (Guo et al., 2009). Significant and dose-related increases of SOD and CAT activities were observed in leaves of plants growing in Elizabeths' river basin in Wyoming, USA contaminated mainly by Cd, Cu, Hg and Zn (Dazy et al., 2009). Similarly Mobin and Khan (2007) observed CAT and SOD activity increase in Cd treated *Brassica juncea*. The increase of SOD, APX and CAT activities in *Phragmites australis* treated with Cd was observed in all parts of the plant (roots, stolons and leaves) in spite of the fact that most of the cadmium was accumulated in roots (Iannelli et al., 2002).

However, it is important to point out that under extreme conditions of stress, a plant may be too weak to produce enough antioxidant enzymes to protect itself. Therefore, low activity of antioxidant enzyme may not always indicate the low stress level and it should be always connected with the evaluation of plant biomass production (Fayiga et al., 2004).

4.5. ANTIOXIDANTS

Except of antioxidant enzymes also other compounds may protect plants against the oxidative stress. Glutathione, phytochelatins, flavonoids, tocoferol and ascorbate belong among such ones. The measurement of their contents in plants can also be used as an indicator of heavy metal induced stress (Bharagava et al. 2008).

Glutathione (GSH) represents one of the major sources of non-protein thiols in most plant cells. A thiol group is important in the formation of mercaptide bond with metals. This reactivity along with the relative stability and high water solubility of GSH makes it an ideal compound to protect plants against stresses including the oxidative stress caused by heavy metal (Foyer, Noctor 2005, Mendoza-Cózatl, Moreno-Sánchez, 2006). Glutathione is also redox-buffer, phytochelatin precursor and substrate for keeping the ascorbate in reduced form in the ascorbate-glutathione cycle (Noctor, Foyer, 1998). It was proposed that glutathione redox state is a good biomarker of the cellular redox state and it reflex the severity of the stress conditions (Kranner et al., 2006, Potters et al., 2010). Non-growth inhibiting Cd concentration in growth medium caused the increase of Cd concentration in the leaves of tobacco seedlings resulting in a slight decrease of glutathione which was observed only during first hours of Cd^{2+}-exposure and, by day 2, was completely recovered to control levels (Vögeli-Lange, Wagner, 1996). The total glutathione pool remained relatively unaffected by Zn exposure to *Avicennia marina* plants (Caregnato et al., 2008). The significant decrease of non-protein thiol content in barley leaves treated with 15 μM Cu was probably due to the enhanced thiol transport to the roots where the excess Cu is initially immobilized. Visible toxicity symptoms at 150 μM – 1500 μM Cu treatment corresponded with a drastic increase of these thiols (Demirevska-Kepova, et al., 2004). The amount of total glutathione and GSH in Zn treated *Phaseolus vulgaris* was significantly higher at the beginning of the experiment, then decreased, but the GSSG content increased slowly during the experiment. The GSSG/GSH ratio was higher but its enhancement was not significant (Cuypers et al., 2001). The GSH concentration increase was observed in *Sedum alfredii* plants treated with Pb. Authors of this study suggested that GSH played an important role in Pb chelatation and detoxification (Gupta et al., 2010). GSH in *Sedum alfredii* increased also in the presence of Cd (Sun et al., 2007). Similarly, the high content of GSH was found in plants *Thlaspi caerulescens* and *Thlaspi praecox* growing on Cd and Zn contaminated soil (Pongrac et al., 2009) and the seeds of bean treated with Cd (Szőllősi et al., 2009). But it was reported that the increase in GSH content above a control level does not always influence metal tolerance in plants but the decrease below this level may weaken the tolerance (Ohlsson et al., 2008). In the study of Devi and Prasat (1998), the level of antioxidants (GSH and other non-protein thiols) exhibited the varied response to the Cu treatments depending on the metal concentration in *Ceratophyllum demersum*. Cu at 2 μM significantly increased (48%) while 4 μM Cu significantly decreased (41%) the GSH content, possibly due to the greater oxidation in the presence of Cu compared to the rate of synthesis. The content of antioxidants such as glutathione and oxidized glutathione significantly increased in the leaves of *Hydrilla verticillata* plants exposed to Zn^{2+}

compared with the control. The ratio of GSSG/GSH increased obviously in plants exposed to 30 mg.l^{-1} Zn^{2+}. The increased ratio of GSSG/GSH suggested that the cellular redox status changed to a more oxidized form though these antioxidant responses were activated in the cell of *Hydrilla verticillata* leaves under the Zn stress. These results suggested that Zn induced oxidative damage and the accelerated operation of antioxidant reactions might provide *Hydrilla verticillata* with the enhanced Zn-stress tolerance (Wang et al., 2009). The total glutathione contents of the roots and shoots of pigeonpea (*Cajanus cajan*) decreased with increasing concentrations of externally supplied Zn or Ni and showed a positive correlation with dry matter accumulation. A high ratio of GSH/GSSG has been maintained (Rao, Sresty, 2000). In plants *Salvinia natans* exposition to a Cr-rich wastewater caused significant enhancement of GSH level (Dhir et al., 2009). Aquatic moss *Fontinalis antipyretica* exposed to Cr^{3+} and Cr^{6+} showed different responses to these two metallic forms. $Cr(NO_3)_3$ in low concentrations caused the decrease of GSH but at higher concentration the significant increase was observed. When $K_2Cr_2O_7$ was applied the decrease of GSH was observed irrespectively of Cr concentration (Dazy et al., 2008). External addition of GSH to *Pteris vittata* an arsenic hyperaccumulator lead to the increase of As uptake by the plant (Wei et al., 2010).

The elevated level of ASC content was found in Cu and Mn treated *Hordeum vulgare* but it was mainly affected by Mn excess (Demirevska-Kepova et al., 2004). The decrease of ASC content together with GSH in plant *Pisum sativum* treated with Cd suggests that the functionality of the ASC–GSH cycle could be limited, and this could explain the increase of the H_2O_2 level observed in pea leaves under cadmium treatment (Romero-Puertas et al., 2007). The ascorbate content decreased significantly in *B. juncea* cultivars with increasing cadmium concentrations (Qadir et al., 2004). The level of ascorbate was not affected by Cr^{6+}, while an increase in the level of glutathione in developing rice seedlings was observed (Panda 2007). The results of Martínez-Domínguez et al. (2008) indicate that *Spartina densiflora* is endowed with an efficient antioxidant system that is rapidly modulated in response to variable stressor levels in soil. The capacity of *Spartina densiflora* to invade and compete successfully in polluted marshes was explained by the ability of the plant to up-regulate the ascorbate cycle and related enzymes. This showed that "relaxing" of antioxidant defense mechanisms, after removing pollutants, is a slow and gradual process (taking weeks), whereas its "activation", after exposure to pollutants, is a rapid process (hours or days), and it could be related to different rates of *de novo* ASC biosynthesis, ASC depletion by ROS scavenging and its recycling from oxidized forms of ASC.

Another big group of antioxidants represents phenolic compounds including α-tocopherol, flavonoids and phenolic acids.

α-Tocopherol, lipid-soluble antioxidant, in cooperation with other antioxidants, is a part of system reducing ROS levels (mainly $^1O_2$ and $OH^\bullet$) in photosynthetic membranes and limiting the extent of lipid peroxidation. α-Tocopherol can physically quench, and therefore deactivate $^1O_2$ in chloroplasts (Munné-Bosch, 2005). The role of α-tocopherol in the reaction of plants to the heavy metal-evoked stress is little known (Gajewska, Skłodowska, 2007)

The increase of α-tocopherol in plant *Lemna minor* in the presence of Cd and Zn was induced. α-Tocopherol along with ASC, GSH seemed to be able partly alleviate the adverse effects of Zn but not of Cd. The pattern of antioxidant response was different when comparing Cd and Zn treatments, with a much larger accumulation of ASC in Cd-treated plants, and GSH in Zn-treated plants (Artetxe et al., 2002). Exposure of the wheat seedlings to Ni led to the enhancement of tocopherol concentration in the shoots. An increase in tocopherol content in the shoots was associated with the enhancement of lipid peroxidation, which indicates that in wheat exposed to Ni, tocopherol may be involved in the protection of the shoot tissues against the oxidative stress. This antioxidant is able to scavenge lipid peroxides, and thus, its increased concentration may result in the reduction of oxidative injuries of cell membranes (Gajewska, Skłodowska, 2007).

Flavonoids occur commonly in leaves, flowering issues, and pollens. They are also abundant in woody parts such as stems and barks. The synthesis of many flavonoids and other phenolic compounds is greatly affected by light. Plants grown in full sun have also been shown to contain higher levels of flavonoids than shade-grown plants (Larson, 1988). In plants flavonoids play an important role as flower pigments (Gidda, Varin, 2006). They are involved in many processes, including plant-pathogen interactions, pollination, and seed development. Flavonoids have been suggested to act as antioxidants, protecting plants from the oxidative stress (Hernández et al., 2008). In spinach H_2O_2 oxidation by flavonol glycosides by intact chloroplasts was evidenced (Takahama, 1984).

Other compounds with antioxidant activity such as leucasin from *Leucas aspera* might still be found (Meghashri et al., 2010).

4.6. PHYTOCHELATINS

Phytochelatins (PCs) are a set of heavy metal binding peptides. They play an important role in heavy metal detoxification as well as in the maintenance of ionic homeostasis. PCs are synthesized inductively by exposure to heavy metals such as Hg, Cu, Zn, Pb and Ni. During the exposure of plants to such metals PCs are synthesized from GSH (Yadav, 2010). The increase of phytochelatin content in maize seedling growing in media with different heavy metals depended on the kind of metal. Its formation was stimulated most effectively by Cd, less by Zn and Cu and negligibly by Ni. Total glutathione declined with Cd and Zn exposure, however, with excess Cu the roots contained 45% more total glutathione than the controls did. The reactions to excess Cd differed along the length of roots. In the 1 cm apical region a high production of PCs occurred with a moderate loss of total glutathione. In the mature region, PC content was 2.5-fold less than in apices, several unidentified thiols accumulated, and total glutathione levels declined drastically (Tukendorf, Rauser, 1990). The concentration of PCs in *Brassica juncea* plants treated with Cd and Cd-EDTA increased significantly during first days of treatment but then sharply decreased. The enhanced level of PCs indicates the plant capacity to detoxify the metal via chelation and sequestration in vacuoles. At higher Cd concentrations after almost one month exposure, plants experienced toxicity besides accumulation of PCs, probably due to GSH depletion (Seth et al., 2008). In the presence of Cd and Pb synthesis of phytochelatins was observed in *Phaeodactylum tricornutum*. The increase of PCs concentration was followed by the consumption of GSH in cells. Over 50% of the cellular glutathione was converted into phytochelatins after a 2-hour- exposure (Morelli, Scarano, 2001). Similarly, the increase of PCs concentration was observed in wheat seedlings cultivated in medium with Cd (Lindberg et al., 2007). Application of heavy metals to cell suspension cultures and whole plants of *Silene vulgaris* and tomato induced the formation of heavy metal–phytochelatin-complexes with Cu and Cd and the binding of Zn and Pb to lower molecular weight substances (Leopold et al., 1999). Srivastava et al. (2006) found that PC concentration positively correlates with no visible toxicity symptomps in *Hydrilla verticillata* exposed to copper. Although PCs were induced significantly by copper in *Hydrilla verticillata*, copper chelation was found to be low suggesting that PCs play only a part in integrated mechanisms of copper homeostasis and detoxification. Visible toxicity symptoms were found in plants which did not synthesize phytochelatins.

However, phytochelatin biosynthesis is apparently not necessary for metal tolerance in *Salix*, as PC was not found in this system either in tolerant or sensitive clones, before or after metal treatment (Ohlsson et al., 2008). No PCs were detected also in plant *Sedum alfredii* treated with Pb and Cd although a significant amount of Pb was accumulated in roots and shoots of these plants simultaneously GSH content increased suggesting that GSH directly might serve as an antioxidant or a metal chelator (Sun et al., 2007, Gupta et al., 2010). On contrary to recent papers reporting that no PCs could be detected in the hyperaccumulator *Sedum alfredi*, in the study of Zhang et al. (2008) PC formation could be induced in the leaf, stem and root tissues of *S. alfredii* which was collected at an old Pb/Zn mining area upon exposure to 400 µM cadmium, and only in the stem and root when exposed to 700 µM lead. However, no PCs were found in any part of *Sedum alfredii* when it was subjected to exposure to 1600 µM zinc. Possible explanation of the differences between the results and those from Sun et al. (2007) relates with the protocol of sample preparation.

In soybean (*Glycine max*) and white lupin (*Lupinus albus*) exposed to Cd and arsenate higher levels of homophytochelatins (hPCs) in soybean and PC levels in white lupin were detected. Higher levels of hPCs in Cd treated soybeans compared to PCs in lupins did not prevent growth inhibition (Vázquez et al., 2009). Cadmium treatment during cultivation of *Triticum aestivum* induced a high amount of PCs, both in root and leaf protoplasts. After inhibition of GSH synthesis with buthionine sulfoximine (BSO) the amount of GSH of both root and leaf protoplasts decreased by approximately 75% as well as the cadmium-induced amount of PCs followed by the decrease of short-term uptake of cadmium into wheat protoplasts (Lindberg et al., 2007). The synthesis of PCs in a marine alga, *Dunalliela tertiolecta*, was strongly induced by Zn. When algal cells were treated with Zn and H_2O_2, a dramatic decrease of PCs but almost no changes in GSH level were observed suggesting that PCs are stronger scavengers of hydrogen peroxide and superoxide radical than glutathione (Tsuji et al., 2002). Arsenate exposure led to PC induction in roots of *Oenothera odorata* while GSH decreased. The presence of PC molecules in *Oenother odorata* roots was found to be a good indicator of As tolerance (Kim et al., 2009). Stolt et al. (2003) have studied PC accumulation in 12-day-old seedlings of two cultivars of spring bread wheat (*Triticum aestivum*), and two spring durum wheat cultivars (*Triticum turgidum var. durum*) with different degrees of Cd accumulation in the grains. They found that high root Cd accumulating cultivars had the highest PC accumulation. Interestingly, these cultivars also exhibited the lowest shoot-to-root Cd concentration ratios, which could be taken to suggest that PC synthesis would contribute, by means of vacuolar sequestration of Cd–PC, to the retention of Cd in the root system. In aquatic fern *Salvinia minima* the accumulation of PC in the plants is a direct response to Pb^{2+} accumulation, and phytochelatins do participate as one of the mechanism to cope with Pb^{2+} of this Pb-hyperaccumulator aquatic fern (Estrella-Gómez et al., 2009).

Chapter 5

CONCLUSION

In summary, phytoremediation processes hold great promise as means of cleaning-up polluted soils and water. But to widen the technology better understanding of heavy metal-induced stress is necessary. Studying heavy metal-plant interactions should improve our understanding of the mechanisms of ion uptake, accumulation and resistance. The evaluation of plant stress in the presence of heavy metals is a very important part of the research with the direct implication to phytoremediation but also plant physiology. Symptoms of stress in plants depend on the particular metal (yet metal combination can act differently), plant species, but also on preliminary adaptation and other factors. Several parameters can be used to assess the heavy metal-induced stress.

Germination is sensitive not only to the kind of metal but also to metal form. The kind of anion can change the toxicity symptoms typical for studied metal. In general, nitrates alleviate the metal toxicity to the seed germination. Metals forming oxoanions in aqueous solutions have detrimental effect on germination irrespectively of present cation.

Metal form does not influence **plant growth** and **biomass production** as significantly as it is in the case of germination. Although some positive effects of nitrates were observed as well. The influence of metal on plant growth and biomass production importantly depends on the plant species and metal concentration.

The level of **photosynthetic pigments** depends not only on the kind of metal and its concentration but also on the time of exposure to metal (or time when the measurement is carried out after the metal exposition). Again, at low metal concentrations no, or a very slight increase of the pigment contents can be observed but high metal concentrations are usually responsible for their decrease. Chlorophylls seem to be more sensitive to metals than carotenoids. The chlorophyll *a*/chlorophyll *b* ratio generally changes in dependence on the kind of metal and plant species. The measurement of chlorophyll *a* fluorescence can give a better picture about the chlorophyll as well as status of the photosynthetic apparatus than only the measurement of the chlorophyll content because heavy metals forming stable substituted chlorophylls do not change the chlorophyll content but exhibit a shift of an absorbance maximum in the photosynthetic pigment extract.

The **enzyme activity** levels can, but need not, change in the presence of metals; the observed changes are usually plant and metal specific. Different metals can have various effects on the same plant species which can differ also with the increase of concentrations of the studied metals. Even in the same plant different levels in enzyme activity can be measured in different plant organs. Changes in enzyme activity are often observed in plants exposed to heavy metals but it is difficult to generalize them. Instead of only one enzyme activity measurement of several enzyme activities measured together would provide more specific picture about the extent of oxidative stress at the cell level. To evaluate the stress for plant even the combination of these measurements along with the measurement of other parameters is necessary.

Among **non-enzymatic antioxidants** GSH and ASC are measured the most often. Usually changes in their levels are identified in the stress conditions but their decrease or increase can be observed depending on the kind of metal, its concentration and plant species. The redox state of GSH can serve as a good biomarker of the cellular redox state and it reflex the severity of the stress conditions. The activation of antioxidant systems is commonly very fast so time of the measurement is also very important. For the generalization it would be necessary to gain more knowledge about the changes of the enzyme activities and antioxidant levels in the plant cells separately for tolerant and sensitive species, young and old plants but especially about time course of changes of the studied compounds quantities. The decrease of the GSH/GSSG ratio (or the increase of the GSSG/GSH ratio) reflects lower capacity of cells to cope with the oxidative stress. The decrease of GSH level may weaken the plant tolerance.

Phytochelatin production by plants in the presence of metals is also not uniform. In some plant species significant increase of the phytochelatin production is found whereas in other no phytochelatin production is detected irrespectively of the plant ability to tolerate elevated metal concentration in the environment. Methodology of the phytochelatin content measurement influence the results markedly as the same metal was found to cause phytochelatin synthesis and not to cause by two different authors studying the same plant.

In regard to differences in plant responses to heavy metal-induced stress especially in the state of antioxidant systems it is necessary to prepare a standard protocol for the evaluation of the impact metals have on plant health but generally it is important for stress evaluation to assess several "stress indicators" together. Combination of parameters expressing the influence on photosynthesis and cell oxidative status would be desirable.

ACKNOWLEDGEMENT

This work was supported by Slovak grant agency – project VEGA1/0134/09.

REFERENCES

Abedin, Md. J., Meharg, A.A (2002). Relative toxicity of arsenite and arsenate on germination and early seedling growth of rice (Oryza sativa L.). *Plant and Soil,* 243, 57-66.

Abhilash, P.C., Jamil, S., Singh, N. (2009). Transgenic plants for enhanced biodegradation and phytoremediation of organic xenobiotics. *Biotechnology Advances,* 27, 474-488.

Abou-Shanab, R.A.I., Angle, J.S., Chaney, R.L. (2006). Bacterial inoculants affecting nickel uptake by *Alyssum murale* from low, moderate and high Ni soils. *Soil Biology and Biochemistry,* 38, 2882-2889.

Ahsan, N., Lee, D.-G., Lee, S.-H., Kang, K.Y., Lee, J. J., Kim, P. J., Yoon, H.-S., Kim, J.-S., Lee, B.-H. (2007). Excess copper induced physiological and proteomic changes in germinating rice seeds. *Chemosphere,* 67, 1182-1193.

Akeredolu, F., Barrie, L.A., Olson, M.P., Oikawa, K.K., Pacyna, J.M. (1994). The flux of anthropogenic trace metals into the Arctic from mid-latitudes in 1989/90. *Atmospheric Environment,* 28, 8, 1557-1572.

Alkorta, I., Garbisu, C. (2001). Phytoremediation of organic contaminants in soil. *Bioresource Technology,* 79, 273-276.

Alvarenga, P., Gonçalves, A.P., Fernandes, R.M., de Varennes, A., Vallini, G., Duarte, E., Cunha-Queda, A.C. (2008). Evaluation of composts and liming materials in the phytostabilization of a mine soil using perennial ryegrass. *Science of the total environment,* 406, 43-56.

An, Y.J. (2006). Assessment of comparative toxicities of lead and copper using plant assay. *Chemosphere,* 62, 8, 1359-1365.

Anderson, M.E. (1998). Glutathione: an overview of biosynthesis and modulation. *Chemico-Biological Interactions,* 111-112, 1-14.

Anderson, C.W.N., Brooks, R.R., Chiaruci, A., LaCoste, C.J., Leblanc, M., Robinson, B.H., Simcock, R., Stewart, R.B. (1999). Phytomining for nickel, thallium and gold. *Journal of Geochemical Exploration,* 67, 407-415.

Anderson, C., Moreno, F., Meech, J. (2005). A field demonstration of gold phytoextraction technology. Minerals Engineering, 18, 385-392.

Apel, K., Hirt, H. (2004). Reactive Oxygen Species: Metabolism, Oxidative Stress, and Signal Transduction. *Annual Review of Plant Biology,* 55, 373-399.

Appenroth, K.-J., Krech, K., Keresztes, Á., Fischer, W., Koloczek, H.(2010). Effects of nickel on the chloroplasts of the duckweeds *Spirodela polyrhiza* and *Lemna minor* and their possible use in biomonitoring and phytoremediation. *Chemosphere,* 78, 3, 216-223.

Aravind, P., Prasad, M. N. V. (2003). Zinc alleviates cadmium-induced oxidative stress in *Ceratophyllum demersum* L.: a free floating freshwater macrophyte. *Plant Physiology and Biochemistry*, 41, 391-397.

Aravind, P. Prasad, M. N. V. (2004). Zinc protects chloroplasts and associated photochemical functions in cadmium exposed *Ceratophyllum demersum* L., a freshwater macrophyte. *Plant Science,* 166, 5, 1321-1327.

Arduini, I., Masoni, A., Mariotti, M., Ercoli, L. (2004). Low cadmium application increase miscanthus growth and cadmium translocation. *Environmental and Experimental Botany*, 52, 89-100.

Arduini, I., Masoni, A., Ercoli, L. (2006). Effects of high chromium applications on miscanthus during the period of maximum growth. *Environmental and Experimental Botany,* 58, 1-3, 234-243.

Arienzo, M., Adamo, P., Cozzolino, V. (2004). The potential of *Lolium perenne* for revegetation of contaminated soils from a metallurgical site. *The Science of the Total Environment,* 319, 13-25.

Arshad, M., Silvestre, J. Pinelli, E., Kallerhoff, J., Kaemmerer, M., Tarigo, A., Shahid, M., Guiresse, M., Pradere, P., Dumat, C. (2008). A field study of lead phytoextraction by various scented *Pelargonium* cultivars. *Chemosphere,* 71, 2187-2192.

Artetxe, U., García-Plazaola, J.-I., Hernández, A., Becerril, J.M. (2002). Low light grown duckweed plants are more protected against the toxicity induced by Zn and Cd. *Plant Physiology and Biochemistry,* 40, 10, 859-863.

Ashraf, M., Harris, PJ.C. (2004). Potential biochemical indicators of salinity tolerance in plants. *Plant Science,* 166, 3-16.

Atal, N., Saradhl, P.P., Mohanty, P. (1991) Inhibition of the chloroplast photochemical reactions by treatment of wheat seedlings with low concentrations of cadmium: analysis of electron transport activities and changes in fluorescence yield. *Plat and Cell Physiology,* 32, 7, 943-951.

Azizur Rahman, M., Hasegawa, H., Mahfuzur Rahman, M., Nazrul Islam, M., Majid Miah, M.A., Tasmen, A. (2007). Effect of arsenic on photosynthesis, growth and yield of five widely cultivated rice (*Oryza sativa* L.) varieties in Bangladesh. *Chemosphere,* 67, 1072-1079.

Baccouch, S., Chaoui, A., El Ferjani, E. (1998). Nickel-induced oxidative damage and antioxidant responses in *Zea mays* shoots, *Plant Physiology and Biochemistry,* 36, 689- 694.

Bajguz, A., Hayat, S. (2009). Effects of brassinosteroids on the plant responses to environmental stresses. *Plant Physiology and Biochemistry,* 47, 1-8.

Baker, A.J.M., Brooks, R.R. (1989). Terrestrial higher plants which hyperaccumulate metallic elements. A review of their distribution, ecology and phytochemistry. *Biorecovery,* 1, 2, 81-126.

Baker, A.J.M., McGrath, S.P., Reeves, R.D., Smith, J.A.C. (2000). Metal hyperaccumulator plants: a review of the ecology and physiology of a biological resource for phytoremediation of metal-polluted soils, Phytoremediation of contaminated soil and water (Terry N, Banuelos G, eds). USA7 Lewis Publishers.

Bali, R., Siegele, R., Harris, A.T. (2010). Phytoextraction of Au: Uptake, accumulation and cellular distribution in *Medicago sativa* and *Brassica juncea*. *Chemical Engineering Journal,* 156, 286-297.

Banásová, V. (1996). Rastliny na substrátoch s vysokým obsahom ťažkých kovov. Zborník zo seminára: *Ťažké kovy v ekosystéme, E'96, BIJO Slovensko, s r.o.*, 81-94.

Bang, S.S, Pazirandeh, M. (1999). Biochemical characteristics of heavy metal uptake by *Escherichia coli* NCP immobilized in kappa-carrageenan beads. Biohydrometallurgy and the Environment toward the Mining of the 21st Century, Proceedings of the International Biohydrometallurgy Symposium IBS´99, Ed.: R.Amils, A.Balester, Elsevier, 193-199.

Bani, A., Echevarria, G., Sulçe, S., Morel, J.L., Mullai, A. (2007). In-situ phytoextraction of Ni by a native population of *Alyssum murale* on an ultramafic site (Albania). *Plant Soil,* 239, 79-89.

Bansal, P., Sharma P., Goyal, V. (2002). Impact of lead and cadmium on enzyme of citric acid cycle in germinating pea seeds. *Biologia Plantarum,* 45, 125-127.

Barbaroux, R., Meunier, N., Mercier, G., Taillard, V., Morel, J.L., Simonnot, M.O., Blais, J.F. (2009). Chemical leaching of nickel from the seeds of the metal hyperaccumulator plant *Alyssum murale. Hydrometallurgy,* 100, 10-14.

Barona, A., Aranguiz, I., Elías, A. (2001). Metal assotiations in soil before and after EDTA extractive decontamination: implications for the effetiones of further clean-up procedures. *Environmental Pollution,* 113, 79-85.

Beceiro-González, E., Taboada-de la Calzada, A., Alonso-Rodrígez, E., López-Mahía, P., Muniategui-Lorenzo, S., Prada-Rodrígez, D. (2000). Interaction between metallic species and biological substrates: approximation to possible intraction Trends in Analytical Chemistry, mechanisms between the alga *Chlorella vulgaris* and arsenic(III). Trends in Analytical Chemistry, 19, 8, 475-480.

Begonia, G.B., Davis, C.D., Begonia, M.F.T., Gray, C.N. (1998). Growth Responses of Indian Mustard [*Brassica juncea* (L.) Czern.] and its Phytoextraction of Lead from a Contaminated Soil, *Bulletin of Environmental Contamination Toxicology,* 61, 38-43.

Bharagava, R.N., Chandra, R., Rai, V. (2008). Phytoextraction of trace elements and physiological changes in Indian mustard plants (*Brassica nigra* L.) grown in post methanated distillery effluent (PMDE) irrigated soil. *Bioresource Technology,* 99, 8316-8324.

Bláha, L., Bocková, R., Hnilička, F., Hniličková, H., Holubec, V., Möllerová, J., Štolcová, J., Zieglerová, J. (2003). *Rostlina a stres* (Plant and Stress), VÚRV, Praha.

Blaudez, D., Kohlerb, A., Martin, F., Sanders, D., Chalot, M. (2003). Poplar Metal Tolerance Protein 1 Confers Zinc Tolerance and Is an Oligomeric Vacuolar Zinc Transporter with an Essential Leucine Zipper Motif. *The Plant Cell,* 15, 2911-2928.

Blaylock, M.J. (2000). Field demonstrations of phytoremediation of lead contaminated soils, Phytoremediation of Contaminated Soil and Water (Terry, N., Bañuelos, G. – eds.), Lewis Publishers, Boca Raton, Florida, 1-12.

Bohnert, H. J., Sheveleva, E. (1998). Plant stress adaptations - making metabolism move. *Current Opinion in Plant Biology,* 1, 267-274.

Bontidean, I., Ahlqvist, J., Mulchandani, A., Chen, W., Bae, W., Mehra, R.K., Mortari, A., Csöregi, E. (2003). Novel synthetic phytochelatin-based capacitive biosensor for heavy metal ion detection. *Biosensors and Bioelectronics,* 18, 547-553.

Boojar, M.M.A., Goodarzi, F. (2007) The copper tolerance strategies and the role of antioxidant enzymes in three plant species grown on copper mine. *Chemosphere, 67,* 2138-2147.

Boominathan, R., Doran, P.M. (2003). Cadmium Tolerance and Antioxidant Defenses in Hairy Roots of the Cadmium Hyperaccumulator, *Thlaspi caerulescens. Biotechnology and Bioengineering, 83,* 158-167.

Boominathan, R., Sada-Chaudhury, N.M., Sahajwalla, V., Doran, P.M. (2004). Production of Nickel Bio-Ore from Hyperaccumulator Plant Biomass: Applications in Phytomining. *Biotechnology and Bioengineering,* 86, 3, 243-250.

Boyd, R.S., Martens, S.N. (1998). The significance of metal hyperaccumulation for biotic interaction. *Chemoecology, 8,* 1-7.

Briat, J.-F., Lebrun, M. (1999). Plant responses to metal toxicity. *Plant biology and pathology, 322,* 43-54.

Broadhurst, C.L., Chaney, R.L., Angle, J.S., Erbe, E.F., Maugel, T.K. (2004). Nickel Localization and Response to Increasing Ni Soil Levels in Leaves of the Ni Hyperaccumulator *Alyssum murale. Plant and Soil,* 265, 225-242.

Brooks, R.R., Chambers, M.F., Nicks, L.J., Robinson, B.H. (1998). Phytomining. *Trends in Plant Science,* 3, 9, 359-362.

Brooks, R.R., Lee, J., Reeves, R.D., Jaffre, T. (1977). Detection of nickeliferous rocks by analysis of herbarium speciments of indicator plants. *Journal of Geochemical Exploration,* 7, 49-57.

Brown, S., Sprenger, M., Maxemchuk, A., Compton, H. (2005). Ecosystem Function in Alluvial Tailings after Biosolids and Lime addition. *Journal of Environmental Quality,* 34, 139-148.

Brun, L.A., Le Corff, J., Maillet, J. (2003). Effects of elevated soil copper on phenology, growth and reproduction of five ruderal plant species. *Environmental Pollution,* 122, 361-368.

Buchanan, B.B., Gruissem, W., Jones, R.L. (2000). Biochemistry and Molecular Biology of Plants. *Courier Companies,* Inc., 843-845.

Cabañero, F.J., Carvajal, M. (2007). Different cation stresses affect specifically osmotic root hydraulic conductance, involving aquaporins, ATPase and xylem loading of ions in *Capsicum annuum,* L. plants. *Journal of Plant Physiology,* 164, 1300-1310.

Cao, H., Jiang, Y., Chen, J., Zhang, H., Huang, W., Li, L., Zhang, W. (2009). Arsenic accumulation in *Scutellaria baicalensis* Georgi and its effects on plant growth and pharmaceutical components. *Journal of Hazardous Materials,* 171, 1-3, 508-513.

Carbonell-Barrachina, A.A., Burló, J. F., Burgos-Hernkdez, A., López, E., Mataix, J. (1997). The influence of arsenite concentration on arsenic accumulation in tomato and bean plants. *Scientia Horticulturae,* 71, 167-176.

Caregnato, F. F., Koller, C.E., MacFarlane G.R., Moreira J.C.F. (2008). The glutathione antioxidant system as a biomarker suite for the assessment of heavy metal exposure and effect in the grey mangrove, *Avicennia marina* (Forsk.) Vierh. *Marine Pollution Bulletin,* 56, 1119-1127.

Cenkci, S., Ciğerci, I.H., Yıldız, M., Özay, C., Bozdağ, A., Terzi, H. (2010). Lead contamination reduces chlorophyll biosynthesis and genomic template stability in *Brassica rapa* L. *Environmental and Experimental Botany,* 67, 467-473.

Chakravarty B., Srivastava, S. (1992). Toxicity of some heavy metals in vivo and in vitro in *Helianthus annuus. Mutation Research Letters,* 283, 287-294.

Chaney, R.L., Angle, J.S., Broadhurst, C.L., Peters, C.A., Tappero, R.V., Sparks, D.L. (2007). Improved Understanding of Hyperaccumulation Yields Commercial Phytoextraction and Phytomining Technologies. *Journal of Environmental Quality,* 36, 1429-1443.

Chaney, R.L., Malik, M., Li, Y.M., Brown, S.L., Brewer, E.P., Angle, J.S., Baker, A.J.M. (1997). Phytoremediation of soil metals. *Current Opinion in Biotechnology,* 8, 279-284.

Chaoui, A., Mazhoudi, S., Ghorbal, M.H., El Ferjani, E. (1997). Cadmium and zinc induction of lipid peroxidation and effects of antioxidant enzyme activities on beans (*Phaseolus vulgaris* L.). *Plant Science,* 127, 139-147.

Chavan, P.V., Dennett, K.E., Marchand, E.A., Gustin, M.S. (2007). Evaluation of small-scale constructed wetland for water quality and Hg transformation. *Journal of Hazardous Materials,* 149, 543-547.

Chen, H.M., Zheng, C.R., Tu, C., Shen, Z.G. (2000). Chemical methods and phytoremediation of soil contaminated with heavy metals. *Chemosphere,* 41, 229-234.

Chen, B., Roos, P., Zhu, Y.-G., Jakobsen, I. (2008). Arbuscular mycorrhizas contribute to phytostabilization of uranium in uranium mining tailings. *Journal of Environmental Radioactivity,* 99, 801-810.

Cho, U-H., Seo, N-H. (2004). Oxidative stress in *Arabidopsis thaliana* exposed to cadmium is due to hydrogen peroxide accumulation. *Plant Science,* 168, 1, 113-120.

Choo, T.P., Lee, C.K., Low, K.S., Hishamuddin, O. (2006). Accumulation of chromium (VI) from aqueous solutions using water lilies (*Nymphaea spontanea*), *Chemosphere*, 62, 961-967.

Choudhary, M., Jetley, U. K., Khan, M. A., Zutshi, S., Fatma, T. (2007). Effect of heavy metal stress on proline, malondialdehyde, and superoxide dismutase activity in the cyanobacterium *Spirulina platensis*-S5. Ecotoxicology and Environmental Safety, *66, 204-209.*

Clemens, S. (a) (2006). Evolution and function of phytochelatin synthases. *Journal of Plant Physiology,* 163, 319-332.

Clemens, S. (b) (2006) Toxic metal accumulation, responses to exposure and mechanisms of tolerance in plants. *Biochimie*, 88, 11, 1707-1719.

Clemens, S., Palmgren, M.G., Krämer, U. (2002). A long way ahead: understanding and engineering plant metal accumulation. *Trends in plant science,* 7, 7, 309-315.

Clemens, S., Peršoh, D. (2009). Multi-tasking phytochelatin synthases. *Plant Science,* 177, 266-271.

Clemente, R., Walker, D.J., Bernal, M.P. (2005). Uptake of heavy metals and As by Brassica juncea grown in a contaminated soil in Aznalcóllar (Spain): The effect of soil amendments. *Environmental Pollution,* 138, 46-58.

Cobbett, C.S. (2000). Phytochelatins and Their Roles in Heavy Metal Detoxification. *Plant Physiology,* 123, 825-832.

Cunningham, S.D., Ow, D.W. (1996). Promises and Prospects of Phytoremediation. *Plant Physiology,* 110, 715-719.

Cuypers, A., Vangronsveld, J., Clijsters, H. (2001). The redox status of plant cells (AsA and GSH) is sensitive to zinc imposed oxidative stress in roots and primary leaves of *Phaseolus vulgaris. Plant Physiology and Biochemistry,* 39, 657-664.

Dazy, M., Béraud, E., Cotelle, S., Meux, E., Masfaraud, J.-F., Férard, J.-F. (2008). Antioxidant enzyme activities as affected by trivalent and hexavalent chromium species in *Fontinalis antipyretica Hedw. Chemosphere*, 73, 281-290.

Dazy, M., Béraud, E., Cotelle, S., Grévilliot, F., Férard, J.-F., Masfaraud, J.-F. (2009). Changes in plant communities along soil pollution gradients: Responses of leaf antioxidant enzyme activities and phytochelatin contents. *Chemosphere*, 77, 376-383.

de Souza, M.P., Chee, H.N., Terry, N. (1999). Rhizosphere bacteria enhance the accumulation of selenium and mercury in wetland plants. *Planta*, 209, 259-263.

Demirevska-Kepova, K., Simova-Stoilova, L., Stoyanova, Z., Hölzer, R., Feller U. (2004). Biochemical changes in barley plants after excessive supply of copper and manganese. *Environmental and Experimental Botany,* 52, 253-266.

Dercová, K., Makovníková, J., Barančíková, G., Žuffa, J. (2005). Bioremediácia toxických kovov kontaminujúcich vody a pôdy. *Chemické listy,* 99, 682-693.

Dhir, B., Sharmila, P., Saradhi, P. P., Nasim, S.A. (2009). Physiological and antioxidant responses of *Salvinia natans* exposed to chromium-rich wastewater. *Ecotoxicology and Environmental Safety,* 72, 6, 1790-1797.

Di Salvatore, M., Carafa, A. M., Carratù, G. (2008). Assessment of heavy metals phytotoxicity using seed germination and root elongation tests: A comparison of two growth substrates, *Chemosphere,* 73, 1461-1464.

Dimkpa, C.O., Merten, D., Svatoš, A., Büchel, G., Kothe, E. (2009). Metal-induced oxidative stress impacting plant growth in contaminated soil is alleviated by microbial siderophores. *Soil Biology and Biochemistry,* 41, 154-162.

Domažlická, E., Vodičková, H., Mader, P. (1994). Fytochelatiny. *Biologické listy,* 59, 2, 81-92.

Domínguez, M.T., Marañón, T., Murillo, J.M., Schulin, R., Robinson, B.H. (2008). Trace elements accumulation in woody plants of the Guadiamar Valley, SW Spain: A large-scale phytomanagement case study. *Environmental Pollution,* 152, 50-59.

Doumett, S., Lamperi, L., Checchini, L., Azzarello, E., Mugnai, S., Mancuso, S., Petruzzelli, G., Del Bubba, M. (2008). Heavy metal distribution between contaminated soil and Paulownia tomentosa, in a pilot-scale assisted phytoremediation study: Influence of different complexing agents. *Chemosphere,* 72, 1481-1490.

Drążkiewicz, M., Baszyński, T. (2010). Interference of nickel with the photosynthetic apparatus of *Zea mays. Ecotoxicology and Environmental Safety* (article in press).

Duffus, J.H. (2002). "Heavy metals" - A meaningless term? *Pure and Applied Chemistry,* 74, 793-807.

Duman, F., Ozturk, F. (2010). Nickel accumulation and its effect on biomass, protein content and antioxidant enzymes in roots and leaves of watercress (*Nasturtium officinale* R. Br.). *Journal of Environmental Sciences,* 22, 4, 526-532.

Dunne, E.J., Culleton, N., O'Donovan, G., Harrington, R., Olsen, A.E. (2005). An integrated constructed wetland to treat contaminants and nutrients from dairy farmyard dirty water. *Ecological Engineering,* 24, 221-234.

Durand, T.C., Sergeant, K., Planchon, S., Carpin, S., Label, P., Morabito, D., Hausman, J-F., Renaut, J. (2010). Acute metal stress in *Populus tremula* × *P. alba* (717-1B4 genotype): Leaf and cambial proteome changes induced by cadmium^{2+}. *Proteomics,* 10, 349-368.

Eapen, S, Suseelan, K.N., Tivarekar, S., Kotwal, S.A., Mitra, R. (2003). Potential for rhizofiltration of uranium using hairy root cultures of *Brassica juncea* and *Chenopodium amaranticolor. Environmental Research,* 91, 127-33.

Ekmekçi, Y., Tanyolaç, D., Ayhan, B. (2008). Effects of cadmium on antioxidant enzyme and photosynthetic activities in leaves of two maize cultivars. *Journal of Plant Physiology,* 165, 600-611.

Elbaz-Poulichet, F., Dupuy, C., Cruzado, A, Velasquez, Z., Achterberg, E.P., Braungardt, C.B. (2000). Influence of Sorption Processes by Iron Oxides and Algae Fixation on Arsenic and Phosphate Cycle in an Acidic Estuary (Tinto River, Spain). *Water Research,* 34, 12, 3222-3230.

El-Ghamery, A.A., El-Kholy, M.A., El-Yousser, M.A.A. (2003). Evaluation of cytological effects of Zn^{2+} in relation to germination and root growth of *Nigella sativa* L. and *Triticum aestivum* L. *Mutation Research/Genetic Toxicology and Environmental Mutagenesis*, 537, 29-41.

Escuder-Gilabert, L., Martín-Biosca, Y., Sagrado, S., Villanueva-Camañas, R.M., Medina-Hernández, M.J. (2001). Biopartitioning micellar chromatography to predict ecotoxicity. *Analytica Chimica Acta,* 448, 173-185.

Espen, L., Pirovano, L., Cocucci, S. M. (1997). Effects of Ni^{2+} during the early phases of radish (*Raphanus sativus*) seed germination. *Environmental and Experimental Botany,* 38, 187-197.

Estrella-Gómez, N., Mendoza-Cózatl, D., Moreno-Sánchez, R., González-Mendoza, D., Zapata-Pérez, O., Martínez-Hernández, A., Santamaría, J.M. (2009). The Pb-hyperaccumulator aquatic fern *Salvinia minima* Baker, responds to Pb^{2+} by increasing phytochelatins via changes in *SmPCS* expression and in phytochelatin synthase activity. *Aquatic Toxicology,* 91, 4, 320-328.

Faller, P., Kienzler, K., Krieger-Liszkay, A. (2005). Mechanism of Cd^{2+} toxicity: Cd^{2+} inhibits photoactivation of Photosystem II by competitive binding to the essential Ca^{2+} site. *Biochimica et Biophysica Acta,* 1706, 158-164.

Farooqui, A., Kulshreshtha, K., Srivastavaas, K., Singh, N., Farooqui, S.A., Pandey V. Ahmad, K.J. (1995). Photosynthesis, stomatal response and metal accumulation in *Cineraria maritima* L. and *Centauria moschata* L. grown in metal-rich soil. *The Science of the Total Environment,* 164, 203-207.

Fayiga, A.O., Ma, L.Q., Cao, X., Rathinasabapathi, B. (2004). Effects of heavy metals on growth and arsenic accumulation in the arsenic hyperaccumulator *Pteris vittata* L. *Environmental Pollution,* 132, 289-296.

Feng, J. (2005). Plant root responses to three abundant soil minerals: silicon, aluminum and iron. *Critical Reviews in Plant Science,* 24, 267-281.

Feng, J., Shi, Q., Wanga, X., Wei, M., Yang, F., Xu, H. (2010). Silicon supplementation ameliorated the inhibition of photosynthesis and nitrate metabolism by cadmium (Cd) toxicity in *Cucumis sativus* L. *Scientia Horticulturae,* 123, 521-530.

Figueroa, J.A.L., Wrobel, K., Afton, S., Caruso, J.A., Corona, J.F.G., Wrobel, K. (2008). Effect of some heavy metals and soil humic substances on the phytochelatin production in wild plants from silver mine areas of Guanajuato, Mexico. *Chemosphere,* 70, 2084-2091.

Fitz, W.J., Wenzel, W.W. (2002). Arsenic transformations in the soil-rhizosphere-plant system: fundamentals and potential application to phytoremediation. *Journal of biotechnology,* 99, 259-278.

Florence, T. M. (1984). The production of Hydroxyl Radical from Hydrogen peroxide. *Journal of Inorganic Biochemistry,* 22, 221-230.

Förstner, U., Wittmann, G.T.W. (1979). Metal Pollution in the Aquatic Environment. 1st ed., Springer-Verlag, Berlin, New York.

Foyer, C.H., Theodoulou, F.L., Delrot, S. (2001). The functions of inter-and intracellular glutathione transport systems in plants. *Trends in Plant Science,* 6, 10, 486-492.

Foyer, C.H., Noctor, G. (a) (2005). Redox homeostasis and antioxidant signalling: a metabolic link between stress perception and physiological response. *Plant Cell,* 17, 1866-1875.

Foyer, C.H., Noctor, G. (b) (2005). Oxidant and antioxidant signaling in plants: a reevaluation of the concept of oxidative stress in a physiological context. *Plant Cell Environment,* 28, 1056-1071.

Foyer, C.H., Noctor, G. (2009). Redox regulation in photosynthetic organisms: signaling, acclimation, and practical implications. *Antioxidants and Redox Signaling,* 11, 861-905.

Freedman, B. (1989). *Environmental Ecology*, Academic Press, Inc., London, 53-80.

Fujita, M., Fujita, Y., Noutoshi, Y., Takahashi, F., Narusaka, Y., Yamaguchi-Shinozaki, K., Shinozaki, K. (2006). Crosstalk between abiotic and biotic stress responses: a current view from the points of convergence in the stress signaling networks. *Current Opinion in Plant Biology*, 9, 436-442.

Gajewska, E., Skłodowska, M. (2007). Relations between tocopherol, chlorophyll and lipid peroxides contents in shoots of Ni-treated wheat. *Journal of Plant Physiology*, 164, 3, 364-366.

Gambale, F., Bregante, M., Paganetto, A., Magistrelli, P., Martella, L., Sacchi, G.A., Rivetta, A., Cocucci, M. (2001). A pilot phytoremediation system for the decontamination of lead-polluted soils. Phytoremediation and wetlands for remediation of contaminated areas. *Proceedings of the Sixth international in situ and on-site bioremediation symposium* (V. Magar, et al. - eds), San Diego, California, Battelle Press, Columbus, Ohio, 6, 5.

Ganesh, K.S., Baskaran, L., Rajasekaran, S., Sumathi, K., Chidambaram, A.L.A., Sundaramoorthy, P. (2008). Chromium stress induced alterations in biochemical and enzyme metabolism in aquatic and terrestrial plants. *Colloids and Surfaces B: Biointerfaces*, 63, 2, 159-163.

Gardea-Torresdey, J.L., Peralta-Videa, J.R., de la Rosa, G., Parsons, J.G. (2005). Phytoremediation of heavy metals and study of the metal coordination by X-ray absorption spectroscopy, Review. *Coordination Chemistry Reviews*, 249, 1797-1810.

Gardea-Torresdey, J.L., Peralta-Videa, J.R., Montes, M., de la Rosa, G., Corral-Diaz, B. (2004). Bioaccumulation of cadmium, chromium and copper by *Convolvulus arvensis* L.: impact on plant growth and uptake of nutritional elements. *Bioresource Technology*, 92, 229-235.

Gažo, J., Kohout, J., Serátor, M., Šramko, T., Zikmund, M. (1974). Všeobecná a anorganická chémia (General and Inorganic Chemistry), 2nd edition, ALFA, Bratislava.

Gerhardt, K. E., Huang, X.-D., Glick, B. R., Greenberg, B. M. (2009). Phytoremediation and rhizoremediation of organic soil contaminants: Potential and challenges. *Plant Science*, 176, 20-30.

Ghnaya, A.B., Charles, G., Hourmant, A., Hamita, J.B., Branchard, M. (2009). Physiological behaviour of four rapeseed cultivar (*Brassica napus* L.) submitted to metal stress. *Comptes Rendus Biologies*, 332, 363-370.

Gidda, S.K., Varin, L. (2006). Biochemical and molecular characterization of flavonoid 7-sulfotransferase from *Arabidopsis thaliana*. *Plant Physiology and Biochemistry*, 44, 628-636.

Gilbert, M., Wagner, H., Weingart, I., Skotnica, J., Nieber, K., Tauer, G., Bergmann, F., Fischer, H., Wilhelm, C. (2004). A new type of thermoluminometer: A highly sensitive tool in applied photosynthesis research and plant stress physiology. *Journal of Plant Physiology*, 161, 641-651.

Glass, D.J. (2000). Phytoremediation of Toxic Metals: Using Plants to Clean Up the Environment, John Wiley and Sons (Raskin, I., Ensley, B.D. – eds.), New York, 15-32.

Glick, B.R. (2003). Phytoremediation: synergistic use of plants and bacteria to clean up the environment. *Biotechnology Advances*, 21, 383-393.

Grandlic, C.J., Palmer, M.W., Maier, R.M. (2009). Optimization of plant growth-promoting bacteria-assisted phytostabilization of mine tailings. *Soil Biology and Biochemistry*, 41, 1734-1740.

Gray, C.W., Dunham, S.J., Dennis, P.G., Zhao, F.J., McGrath, S.P. (2006). Field evaluation of in situ remediation of a heavy metal contaminated soil using lime and red-mud. *Environmental Pollution*, 142, 530-539.

Grichko, V.P., Glick, B.R. (2001). Ethylene and flooding stress in plants. *Plant Physiology and Biochemistry*, 39, 1-9.

Grill, E., Winnacker, E.-L., Zenk, M.H. (1987). Phytochelatins, a class of heavy-metal-binding peptides from plants, are functionally analogous to metallothioneins. *Proceedings of the National Academy of Sciences*, 84, 439-443.

Grime, J.P. (2001). *Plant Strategies, Vegetation Processes and Ecosystem Properties*, Wiley, Chichester.

Guo, B., Liang, Y., Zhu, Y. (2009). Does salicylic acid regulate antioxidant defense system, cell death, cadmium uptake and partitioning to acquire cadmium tolerance in rice? *Journal of Plant Physiology*, 166, 20-31.

Guo, T.R., Zhang, G.P., Zhang, Y.H. (2007). Physiological changes in barley plants under combined toxicity of aluminum, copper and cadmium. *Colloids and Surfaces B: Biointerfaces*, 57, 2, 82-188.

Gupta, A. K., Sinha, S. (2009). Antioxidant response in sesame plants grown on industrially contaminated soil: Effect on oil yield and tolerance to lipid peroxidation. *Bioresource Technology*, 100, 179-185.

Gupta, D.K., Huang, H.G., Yang, X.E., Razafindrabe, B.H.N., Inouhe, M. (2010). The detoxification of lead in *Sedum alfredii* H. is not related to phytochelatins but the glutathione. *Journal of Hazardous Materials*, 177, 437-444.

Hall, J. L. (2002). Cellular mechanisms for heavy metal detoxification and tolerance. In *Journal of Experimental botany*, 53, 1-11.

Hammer, D., Keller, C., McLaughlin, M.J., Hamon, R.E. (2006). Fixation of metals in soil constituents and potential remobilization by hyperaccumulating and non-hyperaccumulating plants: Results from an isotopic dilution study. *Environmental Pollution,* 143, 407-415.

Harris, A.T., Bali, R. (2008). On the formation and extent of uptake of silver nanoparticles by live plants. *Journal of Nanoparticle Research,* 10, 691-695.

Harris, A.T., Naidoo, K., Nokes, J., Walker, T., Orton, F. (2009). Indicative assessment of the feasibility of Ni and Au phytomining in Australia. *Journal of Cleaner Production,* 17, 194-200.

Hattab, S., Dridi, B., Chouba, L., Kheder, M.B., Bousetta, H. (2009). Photosynthesis and growth responses of pea *Pisum sativum* L. under heavy metals stress. *Journal of Environmental Sciences,* 21, 1552-1556.

Havaux, M. (1998). Carotenoids as membrane stabilizers in chloroplasts. Trends Plant Science, 3, 147-151.

Haverkamp. R.G., Marshall, A.T., van Agterveld, D. (2007). Pick your carats: nanoparticles of gold-silver-copper alloy produced in vivo. *Journal of Nanoparticle Research,* 9, 697-700.

Haverkamp, R.G., Marshall, C.A.T. (2009). The mechanism of metal nanoparticle formation in plants: limits on accumulation. *Journal of Nanoparticle Research,* 11, 1453-1463.

Heldt, H-W. (1997). *Plant Biochemistry and Molecular Biology.* Oxford University Press, 296-297.

Hernández, I., Alegre, L., Van Breusegem, F., Munné-Bosch, S. (2009). How relevant are flavonoids as antioxidants in plants? *Trends in Plants Science,* 14, 125-132.

Hernández, J.A., Olmos, E., Corpas, F.J., Sevilla, F., del Riob, L.A. (1995). Salt induced oxidative stress in chloroplasts of pea plants. *Plant Science,* 105, 151-167.

Hirata, K., Tsuji, N., Miyamoto, K. (2005). Biosynthetic Regulation of Phytochelatins, Heavy Metal-Binding Peptides. *Journal of Bioscience and Bioengineering,* 100, 6, 593-599.

Hornik, M., Pipiska, M., Maresova, J., Augustin, J. (2009). Uptake and Translocation of Metal Complexes in Vascular Plants. *Proceedings of 1st Conference on Biotechnology and Metals,* Kosice, Slovakia, 29-32.

Hronec, O. (1996). Ťažké kovy a ich pohyb v pôdach a rastlinách. Zborník zo seminára: *Ťažké kovy v ekosystéme,* E'96, BIJO Slovensko, s r.o., 41-49.

Iannelli, M. A., Pietrini, F., Fiore, L., Petrilli, L., Massacci, A. (2002). Antioxidant response to cadmium in *Phragmites australis* plants. *Plant Physiology and Biochemistry,* 40, 977-982.

Ike, A., Sriprang, R., Ono, H., Murooka, H., Yamashita, M. (2007). Bioremediation of cadmium contaminated soil using symbiosis between leguminous plant and recombinant rhizobia with the MTL4 and the PCS genes. *Chemosphere,* 66, 1670-1676.

Israr, M., Sahi, S. V. (2006). Antioxidant responses to mercury in the cell cultures of *Sesbania drummondii. Plant Physiology and Biochemistry,* 44, 590-595.

Jansen, M.A.K., Hectors K., O'Brien, N.M., Guisez, Y., Potters, G. (2008). Plant stress and human health: Do human consumers benefit from UV-B acclimated crops? *Plant Science,* 175, 449-458.

Jarvis, M.D., Leung, D.W.M. (2001). Chelated lead transport in *Chamaecytisus proliferus* (L.f.) link *ssp. proliferis var. palmensis* (H.Christ): an ultrastructural study. *Plant Science,* 161, 433-441.

Jia, Y., Tang, S., Wang, R., Ju, X., Ding, Y., Tu, S., Smith, D.L. (2010). Effects of elevated CO_2 on growth, photosynthesis, elemental composition, antioxidant level, and phytochelatin concentration in *Lolium mutiforum* and *Lolium perenne* under Cd stress. *Journal of Hazardous Materials,* 180, 1-3, 384-394.

Jiang, H.M., Yang, J.C., Zhang, J.F. (2007). Effects of external phosphorus on the cell ultrastructure and the chlorophyll content of maize under cadmium and zinc stress. *Environmental Pollution,* 147, 750-756.

Kabata-Pendias, A. (2001). *Trace Elements in Soils and Plants.* Third Edition, CRC Press, 2001.

Kabata-Pendias, A., Pendias, H. (1992). *Trace elements in Soils and Plants,* Boca Raton, CRC Press.

Kadukova, J., Kalogerakis, N. (2007). Lead accumulation from non-saline and saline environment by *Tamarix smyrnesis* Bunge. *European Journal of Soil Biology,* 43, 216-223.

Kadukova, J., Manousaki, E., Kalogerakis, N. (2008). Pb and Cd Accumulation and Phyto - Excretion by Salt Cedar (*Tamarix smyrnensis* Bunge). *International Journal of Phytoremediation,* 1, 10, 2008, 31-46.

Karatas, F., Öbek, E., Kamışlı, F. (2009). Antioxidant capacity of *Lemna gibba* L. exposed to wastewater treatment. *Ecological Engineering,* 35, 8, 1225-1230.

Kavuličova, J., Kadukova, J., Podracky, I., Ivánova, D. (2009). Effect of heavy metals on oxidative stress in *Linum usitatissimum*. *1st Internal Conference on Biotechnology and Metals*, 57-60.

Kayser, A., Wenger, K., Keller, A., Attinger, A., Felix, H.R., Gupta, S.K., Schulin, R. (2000). Enhancement of phytoextraction of Zn, Cd and Cu from calcareous soil: the use of of NTA and sulfur amendments. *Environmental Science and Technology*, 32, 1778-1783.

Kazakou, E., Adamidis, G.C., Baker, A.J.M., Reeves, R.D., Godine, M., Dimitrakopoulos, P.G. (2010). Species adaptation in serpentine soils in Lesbos Island (Greece): metal hyperaccumulation and tolerance. *Plant Soil* (article in press).

Ke, S.-S. (2007). Effects of Copper on the Photosynthesis and Oxidative Metabolism of *Amaranthus tricolor* Seedlings. *Agricultural Sciences in China*, 6, 10, 1182-1192.

Khan, A.G., Kuek, C., Chaudhry, T.M., Khoo, C.S., Hayes, W.J. (2000). Role of plants, mycorrhizae and phytochelators in heavy metal contaminated land remediation. *Chemosphere*, 41, 197-207.

Khan, S., Ahmad, I., Shah, T., Rehman, S., Khaliq, A. (2009). Use of constructed wetland for the removal of heavy metals from industrial wastewater. *Journal of Environmental Management*, 90, 3451-3457.

Khatun, S., Ali, M.B., Hahn, E.-J., Paek, K.-Y. (2008). Copper toxicity in *Withania somnifera*: Growth and antioxidant enzymes responses of *in vitro* grown plants. *Environmental and Experimental Botany*, 64, 279-285.

Kim, D.-Y., Park, H., Lee, S.-H., Koo, N., Kim, J.-G. (2009). Arsenate tolerance mechanism of *Oenothera odorata* from a mine population involves the induction of phytochelatins in roots. *Chemosphere*, 75, 4, 505-512.

Kopittke, P.M., Asher, C.J., Blamey F.P.C., Menzies, N.W. (2009). Toxic effects of Cu^{2+} on growth, nutrition, root morphology, and distribution of Cu in roots of Sabi grass. *Science of the Total Environment*, 407, 16, 4613-4621.

Koppolu, L., Agblevor, F.A., Clements, L.D. (2003). Pyrolysis as a technique for separating heavy metals from hyperaccumulators.Part II: Lab-scale pyrolysis of synthetic hyperaccumulator biomass. *Biomass and Bioenergy*, 25, 651-663.

Kos, B., Leštan, D. (2003). Induced Phytoextraction/Soil Washing of Lead Using Biodegradable Chelate and Permeable Barriers. *Environmental Science and Technology*, 37, 624-629.

Kováčik, J., Klejdus, B., Hedvabny, J., Bačkor, M. (2010). Effect of copper and salicylic acid on phenolic metabolites and free amino acids in *Scenedesmus quadricauda* (Chlorophyceae). *Plant Science,* 178, 3, 703-711.

Kováčik, J., Klejdus, B., Kaduková, J., Bačkor, M. (2009). Physiology of *Matricaria chamomilla* exposed to nickel excess. *Ecotoxicology and Environmental Safety,* 72, 2, 603-609.

Kranner, I., Birtic, S., Anderson, K.M., Pritchard, H.W. (2006). Glutathione half-cell reduction potential: a universal stress marker and modulator of programmed cell death? *Free Radical Biology and Medicine,* 40, 2155-2165.

Krantev, A., Yordanova, R., Janda, T., Szalai, G., Popova, L. (2008). Treatment with salicylic acid decreases the effect of cadmium on photosynthesis in maize plants. *Journal of Plant Physiology,* 165, 920-931.

Kumar, R. (2009). Role of naturally occurring osmolytes in protein folding and stability. *Archives of Biochemistry and Biophysics,* 491, 1-6.

Küpper, H., Küpper, F.C., Spiller, M. (1996). Environmental relevance of heavy metal-substituted chlorophylls using the example of water plants. *Journal of Experimental Botany,* 47, 295, 259-266.

Küpper, H., Spiller, M., Küpper, F.C. (2000). Photometric Method fort he Quantification of Chlorophylls and Their Derivates in Complex Mixtures: Fitting with Gauss-peak Spectra. *Analytical Biochemistry,* 286, 2, 247-256.

Kvesitadze, G., Khatisashvili, G., Sadunishvili, T., Ramsden, J.J. (2006). Biochemical Mechanisms of Detoxification in Higher Plants. Springer-Verlag Berlin Heidelberg.

Labra, M., Gianazza, E., Waitt, R., Eberini, I., Sozzi, A., Regondi, S., Grassi, F., Agradi, E. (2006). *Zea mays* L. protein changes in response to potassium dichromate treatments. *Chemosphere,* 60, 1234-1244.

Laloi, C., Apel, K., Danon, A. (2004). Reactive oxygen signalling: the latest news. *Current Opinion in Plant Biology,* 7, 323-328.

Lamb, A.E., Anderson, C.W.N., Haverkamp, R.G. (2001). The extraction of gold from plants and its applications to phytomining. Chemistry in New Zealand, 3, 1-33.

Larson, R. A. (1988). The antioxidants of higher plants. *Phytochemistry,* 27, 969-978.

Lasat, M.M. (2000). Phytoextraction of Metals from Contaminated Soil: A Review of Plant/Soil/Metal Interaction and Assessment of Pertinent Agronomic Issues. *Journal of Hazardous Substance Research*, 2, 1-25.

Lasat, M.M. (2002). Phytoextraction of toxic metals, a review of biological mechanisms. *Journal of Environmental Quality*, 31, 109-120.

Lebeau, T., Braud, A., Jézéquel, K. (2008). Performance of bioaugmentation-assisted phytoextraction applied to metal contaminated soils: A review. *Environmental Pollution*, 153, 497-522.

Lee, G., Carrow, R.N., Duncan, R.R., Eiteman, M.A., Rieger, M.W. (2008). Synthesis of organic osmolytes and salt tolerance mechanisms in *Paspalum vaginatum*. *Environmental and Experimental Botany*, 63, 19-27.

Lee, S.-H., Ahsan, N., Lee, K.-W., Kim, D.-H., Lee, D.-G., Kwak, S.-S., Kwon, S.-Y., Kim, T.-H., Lee, B.-H. (2007). Simultaneous overexpression of both CuZn superoxide dismutase and ascorbate peroxidase in transgenic tall fescue plants confers increased tolerance to a wide range of abiotic stresses. *Journal of Plant Physiology*, 164, 1626-1638.

Leita, L., Nobili, M. D., Mondini, C., Garcia, M. T. B. (1993). Response of Leguminosae to cadmium exposure. *Journal of Plant Nutrition*, 16, 2001-2012.

León, A. M., Palma, J. M., Corpas, F. J., Gómez, M., Romero-Puertas, M. C., Chatterjee, D., Mateos, R., M., del Río, L. A., Sandalio, L. M. (2002). Antioxidant enzymes in cultivars of pepper plants with different sensitivity to cadmium. *Plant Physiology and Biochemistry*, 40, 813-820.

Leopold, I., Günther, D., Schmidt, J., Neumann, D. (1999). Phytochelatins and heavy metal tolerance. *Phytochemistry*, 50, 1323-1328.

Lewandowski, I., Schmidt, U., Londo, M., Faaij, A. (2006). The economic value of the phytoremediation function – assessed by the example of cadmium remediation by willow (*Salix* ssp.). *Agricultural Systems*, 89, 68-89.

Li, C.-X, Feng, S.-L., Shao, Y., Jiang, L.-N., Lu, X.-Y., Hou, X.-L. (2007). Effects of arsenic on seed germination and physiological activities of wheat seedlings. *Journal of Environmental Sciences*, 19, 725-732.

Li, H.F., Gray, C., Mico, C., Zhao, F.-J., McGrath, S.P. (2009). Phytotoxicity and bioavailability of cobalt to plants in a range of soils. *Chemosphere*, 75, 7, 979-986.

Li, Y.M., Chaney, R.L., Brewer, E.P., Rosenberg, R.J., Angle, J.S., Baker, A.J.M., Reeves, R.D., Nelkin, J. (2003). Development of a technology for commercial phytoextraction of nickel: economic and technical considerations. *Plant Soil,* 249, 107-115.

Li, J.T., Liao, B., Dai, Z.Y., Zhu, R., Shu, W.S. (2009). Phytoextraction of Cd-contaminated soil by carambola (*Averrhoa carambola*) in field trials. *Chemosphere,* 76, 1233-1239.

Lievens, C., Yperman, J., Vangronsveld, J., Carleer, R. (2008). Study of the potential valorisation of heavy metal contaminated biomass via phytoremediation by fast pyrolysis: Part I. Influence of temperature, biomass species and solid heat carrier on the behaviour of heavy metals. *Fuel,* 87, 1894-1905.

Lindberg, S., Landberg, T., Greger, M. (2007). Cadmium uptake and interaction with phytochelatins in wheat protoplasts. *Plant Physiology and Biochemistry,* 45, 47-53.

Liu, X., Zhang, S., Shan, X., Zhu, Y.G. (2005). Toxicity of arsenate and arsenite on germination seedling growth and amylolytic activity of wheat. *Chemosphere,* 61, 293-301.

Liua, Y-J., Mub, Y-J., Zhub, Y-G., Dinga, H., Arens, N.C. (2007). Which ornamental plant species effectively remove benzene from indoor air? *Atmospheric Environment,* 41, 650-654.

Lombardi, L., Sebastiani, L. (2005). Copper toxicity in *Prunus cerasifera*: growth and antioxidant enzymes responses of in vitro grown plants. *Plant Science,* 168, 3, 797-802.

López-Bucio, J., Nieto-Jacobo, M.F., Ramírez-Rodríguez, V., Herrera-Estrella, L. (2000). Organic acid metabolism in plants: from adaptive physiology to transgenic varieties for cultivation in extreme soils. *Plant Science,* 160, 1-13.

MacFarlane, G.R., Burchett, M.D. (2001). Photosynthetic Pigments and Peroxidase Activity as Indicators of Heavy Metal Stress in the Grey Mangrove, *Avicennia marina* (Forsk.) Vierh. *Marine Pollution Bulletin,* 42, 233-240.

Macinnis-Ng, C.M.O., Ralph, P.J. (2002). Towards a more ecologically relevant assessment of the impact of heavy metals on the photosynthesis of the seagrass, *Zostera capricorni. Marine Pollution Bulletin,* 45, 100-106.

Mackova, M., Dowling, D., Macek, T. (2006). Phytoremediation, Rhizoremediation. *Focus on Biotechnology,* 9A, Spinger, Berlin, New York.

Madrid, F., Liphadzi, M.S., Kirkham, M.D. (2003). Heavy metal displacement in chelate-irrigated soil during phytoremediation. *Journal of Hydrology*, 272, 107-119.

Maestri, E., Marmiroli, M., Visioli, G., Marmiroli, N. (2010). Metal tolerance and hyperaccumulation: Costs and trade-offs between traits and environment. *Environmental and Experimental Botany*, 68, 1-13.

Mahajan, S., Tuteja, N. (2005). Cold, salinity and drought stresses: An overview. *Archives of Biochemistry and Biophysics*, 444, 139-158.

Majer Newman, J., Clausen, J.C., Neafsey, J.A. (2000). Seasonal performance of a wetland constructed to process dairy milkhouse wastewater in Connecticut. *Ecological Engineering*, 14, 181-198.

Maleva, M.G., Nekrasova, G.F., Malec, P., Prasad, M.N.V., Strzałka, K. (2009). Ecophysiological tolerance of *Elodea canadensis* to nickel exposure. *Chemosphere*, 77, 3, 392-398.

Malmström, B.G., Leckner, J. (1998). The chemical biology of copper. *Current Opinion in Chemical Biology*, 2, 286-292.

Manousaki, E., Kadukova, J., Papadantonakis, N., Kalogerakis, N. (2008). Phytoextraction and phytoexcretion of Cd by the leaves of *Tamarix smyrnensis* growing on contaminated non saline and saline soils. *Environmental Research*, 106, 3, 326-332.

Marchiol, L., Assolari, S., Sacco, P., Zerbi, G. (2004). Phytoextraction of heavy metals by canola (*Brassica napus*) and radish (*Raphanus sativus*) grown on multicontaminated soil. *Environmental Pollution*. 132, 21-27.

Martínez-Domínguez, D., de las Heras, M.A., Navarro, F., Torronteras, R., Córdoba, F. (2008). Efficiency of antioxidant response in *Spartina densiflora*: An adaptative success in a polluted environment. *Environmental and Experimental Botany*, 62, 1, 69-77.

Masarovičová, E., Repčák, M., Erdelský, K., Gašparíková, O., Ješko, T., Mistrík, I. (2008). Fyziológia rastlín (Plant Physiology), Univerzita Komenského Bratislava.

Mascher, R., Lippmann, B., Holzinger, S., Bergmann, H. (2002.) Arsenate toxicity: effects on oxidative stress response molecules and enzymes in red clover plants. *Plant Science*, 163, 961-969.

Maxted, A.P., Black, C.R., West, H.M., Crout, N.M.J., McGrath, S.P., Young, S.D. (2007). Phytoextraction of cadmium and zinc from arable soils amended with sewage sludge using *Thlaspi caerulescens*: Development of predictive model. *Environmental Pollution*, 150, 363-372.

Mazhoudi, S., Chaoui, A., Ghorbal, M. H., El Ferjani, E. (1997). Response of antioxidant enzymes to excess copper in tomato (*Lycopersicon esculentum*, Mill.). *Plant Science,* 127, 129-137.

McGrath, S.P., Lombi, E., Gray, C.W., Caille, N., Dunham, S.J., Zhao, F.J. (2006). Field evaluation of Cd and Zn phytoextraction potential by the hyperaccumulators *Thlaspi caerulescens* and *Arabidopsis halleri*. *Environmental Pollution,* 141, 1, 115-125.

McGrath, S.P., Zhao, F.J., Lombi, E. (2002). Phytoremediation of metals, metalloids, and radionuclides. *Advances in Agronomy,* 75, 1-56.

Meers, E., Lamsal, S., Vervaeke, P., Hopgood, P., Lust, N., Tack, F.M.G., Verloo, M.G. (2005). Availability of heavy metals for uptake by *Salix viminalis* on a moderately contaminated dredged sediment disposal site. *Environmental Pollution,* 137, 354-364.

Meers, E., Van Slycken, S., Adriaensen, K., Ruttens, A., Vangronsveld, J., Du Laing, G., Witters, N., Thewys, T., Tack, F.M.G. (2010). The use of bio-energy crops (*Zea mays*) for 'phytoattenuation' of heavy metals on moderately contaminated soils: *A field experiment. Chemosphere,* 78, 35-41.

Meghashri, S., Vijay Kumar, H., Gopal, S. (2010). Antioxidant properties of a novel flavonoid from Leaves of *Leucas aspera. Food Chemistry,* 122, 1, 105-110.

Mehrabi, M., Hosseinkhani, S., Ghobadi, S. (2008). Stabilization of firefly luciferase against thermal stress by osmolytes. *International Journal of Biological Macromolecules,* 43, 187-191.

Mench, M., Lepp, N., Bert, V., Schwitzguébel, J.-P., Gawronsi, S.W., Schröder, P., Vangronsveld, J. (2010). Successes and limitations of phytotechnologies at field scale: outcomes, assessment and Outlook from COST Action 859. *Journal of Soils and Sediments* (article in press).

Mendez, M.O., Maier, R.M. (2008). Phytostabilization of mine tailings in arid and semiarid environments - an emerging remediation technology. *Environmental Health Perspectives,* 116, 278-283.

Mendoza-Cózatl, D.G., Moreno-Sánchez, R. (2006). Control of glutathione and phytochelatin synthesis under cadmium stress. Pathway modeling for plants. *Journal of Theoretical Biology,* 238, 919-936.

Milone, M.T., Sgherri, C., Clijsters, H., Navari-Izzo, F. (2003). Antioxidant responses of wheat treated with realistic concentration of cadmium. *Evironmental and experimental botany,* 50, 265-276.

Mishra, S., Srivastava, S., Tripathi, R. D., Kumar, R., Seth, C. S., Gupta, D. K. (2006). Lead detoxification by coontail (*Ceratophyllum demersum* L.)

involves induction of phytochelatins and antioxidant system in response to its accumulation. *Chemosphere,* 65, 1027-1039.

Mithöfer, A., Schulze, B., Boland, W. (2004). Biotic and heavy metal stress response in plants: evidence for common signals. *Federation of European Biochemical Societies Letters,* 566, 1-5.

Mittler, R. (2002). Oxidative stress, antioxidants and stress tolerance. *Trends in Plant Science,* 7, 405-410.

Mleczek, M., Rissmanna, I., Rutkowski, P., Kaczmarek, Z., Golinski, P. (2009). Accumulation of selected heavy metals by different genotypes of *Salix. Environmental and Experimental Botany,* 66, 289-296.

Mobin, M., Khan, N. A. (2007). Photosynthetic activity, pigment composition and antioxidant response of two mustard (*Brassica juncea*) cultivars differing in photosynthetic capacity subjected to cadmium stress. *Journal of Plant Physiology,* 164, 601-610.

Møller, I. M. (2001). Plant mitochondria and oxidative stress: electron transport, NADPH turnover, and metabolism of reactive oxygen species. *Annual Rewiew of Plant Physiology and Plant Molecular Biology,* 52, 561-591.

Monferrán, M.V., Agudo, J.A.S., Pignata, M.L., Wunderlin, D.A. (2009). Copper-induced response of physiological parameters and antioxidant enzymes in the aquatic macrophyte *Potamogeton pusillus. Enviromental pollution,* 157, 2570-2576.

Monteiro, M.S., Santos, C., Soares, A.M.V.M., Mann, R.M. (2009). Assessment of biomarkers of cadmium stress in lettuce. *Ecotoxicology and Enviromental safety,* 72, 811-818.

Morelli, E., Scarano, G. (2001). Synthesis and stability of phytochelatins induced by cadmium and lead in the marine diatom *Phaeodactylum tricornutum, Marine Environmental Research,* 52, 383-395.

Mori, S., Uraguchi, S., Ishikawa, S., Araoa, T. (2009). Xylem loading process is a critical factor in determining Cd accumulation in the shoots of *Solanum melongena* and *Solanum torvum. Environmental and Experimental Botany,* 67, 127-132.

Munné-Bosch, S. (2005). The role of α-tocopherol in plant stress tolerance. *Journal of Plant Physiology,* 162, 7, 743-748.

Munzuroglu, O., Geckil, H. (2002). Effects of Metals on Seed Germination, Root Elongation, and Coleoptile and Hypocotyl Growth in *Triticum aestivum* and *Cucumis sativus. Archives of Environmental Contamination and Toxikology,* 43, 203-213.

Murakami, M., Ae, N. (2009). Potential of phytoextraction of copper, lead and zinc by rice (*Oryza sativa* L.), soybean (*Glycine max* [L.] Merr.), and maize (*Zea mays* L.). *Journal of Hazardous Materials*, 162, 1185-1192.

Murphy, A., Taiz, L. (1995). Comparison of metallothionein gene expression and non-protein thiols in 10 *Arabidopsis* ecotypes. *Plant Physiology*, 109, 945-954.

Nedelkoska, T.V., Doran, P.M. (2000). Characteristics of heavy metal uptake by plant species with potential for phytoremediation and phytomining. *Minerals Engineering*, 13, 5, 549-561.

Neil, S., Desikan, R., Hancock, J. (2002). Hydrogen peroxide signalling. *Current Opinion in Plant Biology*, 5, 388-395.

Nessner Kavamura, V., Esposito, E. (2010). Biotechnological strategies applied to the decontamination of soils polluted with heavy metals. *Biotechnology Advances*, 28, 61-69.

Noctor, G., Foyer, C. H. (1998). Simultaneous Measurement of Foliar Glutathione, γ-Glutamylcysteine, and Amino Acids by High-Performance Liquid Chromatography: Comparison with Two Other Assay Methods for Glutathione. *Analytical Biochemistry*, 264, 98-110.

Ohlsson, A. B., Landberg, T., Berglund, T., Greger, M. (2008). Increased metal tolerance in *Salix* by nicotinamide and nicotinic acid. *Plant Physiology and Biochemistry*, 46, 655-664.

Orcutt, D.M., Nilsen, E.T. (2000). The Physiology of Plants under Stress, Soil and Biotic Factors. JohnWiley&Sons.

Otte, M.L. (2001). What is stress to a wetland plant? *Environmental ans Experimental Botany*, 46, 195-202.

Ouyang, Y. (2002). Phytoremediation: modelling plant uptake and contaminant transport in the soil-plant-atmosphere continuum. *Journal of Hydrology*, 266, 66-82.

Ozturk, F., Duman, F, Leblebici, Z., Temizgul, R. (2010). Arsenic accumulation and biological responses of watercress (*Nasturtium officinale* R. Br.) exposed to arsenite. *Environmental and Experimental Botany*, 69, 2, 167-174.

Palma, J.M., Sandalio, L.M., Corpas, F.J., Romero-Puertas, M.C., McCarthy, I., del Rio, L.A. (2002). Plant proteases, protein degradation, and oxidative stress: role of peroxisomes. *Plant Physiology and Biochemistry*, 40, 521-530.

Panda, S.K. (2007). Chromium-mediated oxidative stress and ultrastructural changes in root cells of developing rice seedlings. *Journal of Plant Physiology*, 164, 11, 1419-1428.

Pandey, N., Sharma, C.P. (2002). Effect of heavy metals Cu^{2+}, Ni^{2+} and Cd^{2+} on growth and metabolism of cabbage. *Plant Science,* 163, 753-758.

Pandey, V., Dixit, V., Shyam, R. (2005). Antioxidant responses in relation to growth of mustard (*Brassica juncea* cv. Pusa Jaikisan) plants exposed to hexavalent chromium. *Chemosphere,* 61, 1, 40-47.

Parvanova, D., Ivanov, S., Konstantinova, T., Karanov, E., Atanassov, A., Tsvetkov, T., Alexieva, V., Djilianov, D. (2004). Transgenic tobacco plants accumulating osmolytes show reduced oxidative damage under freezing stress. *Plant Physiology and Biochemistry,* 42, 57-63.

Pavlíková, D., Pavlík, M., Staszková, L., Motyka, V., Száková, J., Tlustoš, P., Balík, J. (2008). Glutamate kinase as a potential biomarker of heavy metal stress in plants. *Ecotoxicology and Environmental Safety,* 70, 223-230.

Peng, K., Luo, C., Lou, L., Li, X., Shen, Z. (2008). Bioaccumulation of heavy metals by the aquatic plants *Potamogeton pectinatus* L. and *Potamogeton malaianus* Miq. and their potential use for contamination indicators and in wastewater treatment. *Science of the total environment,* 392, 22-29.

Peralta, J.R., Gardea-Torresdey, J.L., Tiemann, K.J., Gomez, E., Arteaga, S., Rascon, E., Parsons, J.G. (2001). Uptake and effects of five heavy metals on seed germination and plant growth in alfalfa (*Medicago sativa* L.). *Bulletin of Environmental Contamination and Toxicology,* 66, 727-734.

Pérez-de-Mora, A., Burgos, P., Cabrera, Madejón, E. (2007). "In Situ" Amendments and Revegetation Reduce Trace Element Leaching in a Contaminated Soil. *Water, Air and Soil Pollution,* 185, 209-222.

Persans, M.W., Salt, D.E. (2000). Possible molecular mechanism involved in nickel, zinc and selenium hyperaccumulation in plants. *Biotechnology* & *Genetic Engineering* Reviews, 17, 389-413.

Pichtel, J., Kuroiwa, K., Sawyerr, H.T. (2000). Distribution of Pb, Cd and Ba in soils and plants of two contaminated sites. *Environmental Pollution,* 110, 171-178.

Piechalak, A., Tomaszewska, B., Baralkiewicz, D. (2003). Enhancing phytoremediative ability of *Pisum sativumby* EDTA application. *Phytochemistry,* 64, 1239-1251.

Pietramellara, G., Nannipieri, P., Renella, G. (2009). Microbial biomass, respiration and enzyme activities after in situ aided phytostabilization of a Pb- and Cu-contaminated soil. *Ecotoxicology and Environmental Safety,* 72, 115-119.

Pitzschke, A., Hirt, H. (2006). Mitogen-Activated protein Kinases and Reactive Oxygen Species Signaling in Plants. *Plant Physiology,* 141, 351-356.

Pongrac, P., Zhao, F.-J., Razinger, J., Trámec, A., Regvar, M. (2009). Physiological responses to Cd and Zn in two Cd/Zn hyperaccumulating *Thlaspi* species. *Environmental and Experimental Botany,* 66, 479-486.

Porra, R.J. (2002). The chequered history of the development and use of simultaneous equations for the accurate determination of chlorophylls *a* and *b*. *Photosynthesis Research,* 73, 149-156.

Posmyk, M. M., Kontek, R., Janas, K.M. (2009). Antioxidant enzymes activity and phenolic compounds content in red cabbage seedlings exposed to copper stress. *Ecotoxicology and Environmental Safety,* 72, 596-602.

Potters, G., Horemans, N., Jansen, M.A.K. (2010). The cellular redox state in plant stress biology – A charging concept. *Plant Physiology and Biochemistry,* 48, 5, 292-300.

Poulik, Z. (1999). Influence of nickel contaminated soils on lettuce and tomatoes *Scientia Horticulturae*, 81, 3, 243-250.

Prado, C., Rodríguez-Montelongo, L., González, J. A., Pagano, E.A., Hilal, M., Prado, F.E. (2010). Uptake of chromium by *Salvinia minima*: Effect on plant growth, leaf respiration and carbohydrate metabolism. *Journal of Hazardous Materials,* 177, 1-3, 546-553.

Prasad, M.N.V., Malec, P., Waloszek, A., Bojko, M., Strzalka, K. (2001). Physiological responses of *Lemna trisulca* L. (duckweed) to cadmium and copper bioaccumulation. *Plant Science,* 161, 881-889.

Procházka, S., Macháčková, I., Krekule, J., Šebánek J. a kol. (1998). Fyziologie rostlin (Plant physiology), Academia, Praha.

Pulford, I.D., Watson, C. (2003). Phytoremediation of heavy metal-contaminated land by trees – a review. *Environment International,* 29, 529-540.

Puschenreiter, M., Stöger, G., Lombi, E., Horak, O., Wenzel, W.W. (2001). Phytoextraction of heavy metal contaminated soil with *Thlaspi goesingense* and *Amaranthus hybridus*: Rhizosphere manipulation using EDTA and ammonium sulfate. *Journal of Plant Nutrition and Soil Science,* 164, 615-621.

Qadir, S., Qureshi, M.I., Javed, S., Abdin, M.Z. (2004). Genotypic variation in phytoremediation potential of *Brassica juncea* cultivars exposed to Cd stress. *Plant Science,* 167, 5, 1171-1181.

Qiu, R.-L., Zhao, X., Tang, Y.-T., Yu, F.-M., Hu, P.-J. (2008). Antioxidant response to Cd in a newly discovered cadmium hyperaccumulator, *Arabis paniculata* F. *Chemosphere,* 74, 6-12.

Quartacci, M.F., Irtelli, B., Baker, A.J.M. Navari-Izzo, F. (2007). The use of NTA and EDDS for enhanced phytoextraction of metals from a multiply contaminated soil by *Brassica carinata*. *Chemosphere*, 68, 1920-1928.

Qureshi, M.I., Qadir, S., Zolla, L. (2007). Proteomics-based dissection of stress-responsive pathways in plants. *Journal of Plant Physiology*, 164, 1239-1260.

Radić, S., Babić, M., Škobić, D., Roje, V., Pevalek-Kozlina, B. (2010). Ecotoxicological effects of aluminum and zinc on growth and antioxidants in *Lemna minor* L. *Ecotoxicology and Environmental Safety*, 73, 336-342.

Rahman, M.A., Hasegawa, H., Rahman, M.M., Islam, M.N., Miah, M.A.M., Tasmen, A. (2007). Effect of arsenic on photosynthesis, growth and yield of five widely cultivated rice (*Oryza sativa* L.) varieties in Bangladesh. *Chemosphere*, 67, 6, 1072-1079.

Rahoui, S., Chaoui, A., Ferjani E. E. (2010). Membrane damage and solute leakage from germinating pea seed under cadmium stress. *Journal of Hazardous Materials*, 178, 1-3, 1128-1131.

Rai, V., Vajpayee, P., Singh, S.N., Mehrotra, S. (2004). Effect of chromium accumulation on photosynthetic pigments, oxidative stress defence systems, nitrate reduction, proline level and eugenol content of *Ocimum tenuiflorum* L. *Plant Science*, 167, 1159-1169.

Rama Devi, S., Prasad, M.N.V. (1998). Copper toxicity in *Ceratophyllum demersum* L. (Coontail), a free floating macrophyte: Response of antioxidant enzymes and antioxidants. *Plant Science*, 138, 2, 157-165.

Rama Devi, S., Prasad, M. N. V. (2004). Membrane Lipid Alterations in Heavy Metal Exposed Plants. In Prasad, M. N. V. Heavy Metal Stress: from biomolecules to Ecosystems in Plants. Berlin, Heidelberg: Springer Verlag, 462.

Rao, K.V.M., Sresty, T.V.S. (2000). Antioxidant parameters in the seedlings of pigeonpea (*Cajanus cajan* (L.) Millspaugh) in response to Zn and Ni stresses. *Plant Science*, 157, 1, 113-128.

Raskin, I., Nanda Kumar, P.B.A., Dushenkov, S., Salt, D.E. (1994). Bioconcentration of Heavy Metals by Plants. *Current Opinion in Biotechnology*, 5, 3, 285-290.

Rauser, W. E. (1999). Structure and Function of Metal Chelators Produced by Plants: The case for organic acids, amino acids, phytin, and metallothioneins. *Cell Biochemistry and Biophysics*, 31, 19-48.

Rhoads, D.M., Umbach, A.L., Subbaiah, C.C., James, N., Siedow, J.N. (2006). Mitochondrial reactive oxygen species. Contribution to oxidative stress and interorganellar signaling. *Plant Physiology*, 141, 357-366.

Rios-Gonzalez, K., Erdei, L., Lips, S.H. (2002). The activity of antioxidant enzymes in maize and sunflower seedlings as affected by salinity and different nitrogen sources. *Plant Science,* 162, 923-930.

Rizzi, L., Petruzelli, G., Poggio, G., Vigna Guidi, G. (2004). Soil physical changes and plant availability of Zn and Pb in a treatability test of phytostabilization. *Chemosphere,* 57, 1039-1046.

Robinson, B.H., Brooks, R.R., Clothier, B.E. (1999). Soil amendments affecting nickel and cobalt uptake by *Berkheya coddii*: potential use for phytomining and phytoremediation. *Annals of Botany,* 84, 689-694.

Robinson, B.H, Chiarucci, A., Brooks, R. R., Petit, D., Kirkman, J.H., Gregg, P.E.H., De Dominicis, V. (1997). The nickel hyperaccumulator plant *Alyssum bertolonii* as a potential agent for phytoremediation and phytomining of nickel. *Journal of Geochemical Exploration,* 59, 2, 75-86

Robinson, B., Fernández, J-E., Madejón, P., Marañón, T., Murillo, J.M., Green, S., Clothier, B. (2003). Phytoextraction: an assessment of biogeochemical and economic viability. *Plant and Soil,* 249, 117-125.

Rodrígez-López, J.N., Espín, J.C., del Amor, F., Tudela, J., Martínez, V., Cerdá, A., García-Cánovas, F. (2000). Purification and Kinetic Characterization of an Anionic Peroxidase from Melon (*Cucumis melo* L.) cultivated under Different Salinity Conditions. *Journal of Agricultural and Food Chemistry,* 48, 537-1541.

Rodriguez, L.M., Alatossava, T. (2010). Effects of copper on germination, growth and sporulation of *Clostridium tyrobutyricum. Food Microbiology,* 27, 3, 434-437.

Romero-Puertas, M.C., Corpas, F. J., Rodríguez-Serrano, M., Gómez, M., del Río, L.A., Sandalio, L.M. (2007). Differential expression and regulation of antioxidant enzymes by cadmium in pea plants. *Journal of Plant Physiology,* 164, 1346-1357.

Römkens, P., Bouwman, L., Japenga, J., Draaisma, C. (2002). Potentials and drawbacks of chelate-enhanced phytoremediation of soils. *Environmental Pollution,* 116, 109-121.

Rooney, C.P., Zhao, F.-J. McGrath, S.P. (2007). Phytotoxicity of nickel in a range of European soils: Influence of soil properties, Ni solubility and speciation. *Environmental Pollution,* 145, 2, 596-605.

Rousseau, D.P.L., Vanrolleghem, P.A., De Pauwa, N. (2004). Model-based design of horizontal subsurface flow constructed treatment wetlands: a review. *Water Research,* 38, 1484-1493.

Rout, G.R., Samantaray, S., Das, P. (2000). Effects of chromium and nickel on germination and growth in tolerant and non-tolerant populations of *Echinochloa colona* (L.) Link. *Chemosphere*, 40, 855-859.

Ruley, A.T., Sharma, N.C., Sahi, S.V. (2004). Antioxidant defense in a lead accumulating plant, *Sesbania drummondii*. *Plant Physiology and Biochemistry*, 42, 899-906.

Ruttens, A., Colpaert, J.V., Mench, M., Boisson, J., Carleer, R., Vangronsveld, J. (a) (2006). Phytostabilization of metal contaminated sandy soil. I: Influence of compost and/or inorganic metal immobilizing soil amendments on metal leaching. *Environmental Pollution*, 144, 533-539.

Ruttens, A., Mench, M., Colpaert, J.V., Boisson, J., Carleer, R., Vangronsveld, J. (b) (2006). Phytostabilization of metal contaminated sandy soil. I: Influence of compost and/or inorganic metal immobilizing soil amendments on phytotoxicity and plant availability of metals. *Environmental Pollution*, 144, 524-532.

Ryser, P., Sauder, W.R. (2006). Effects of heavy-metal-contaminated soil on growth, phenology and biomass turnover of *Hieracium piloselloides*. *Environmental Pollution*, 140, 52-61.

Saifullah, Meers, E., Qadir, M., de Caritat, P., Tack, F.M.G., Du Laing, G., Zia, H. (2009). EDTA-assisted Pb phytoextraction. *Chemosphere*, 74, 1279-1291.

Salt, D.E., Smith, R.D., Raskin, I. (1998). Phytoremediation. *Annual Review of Plant Physiology and Plant Molecular Biology*, 49, 643-668.

Sandalio, L.M., Dalurzo, H.C., Gómez, M., Romero-Puertas, M.C., del Río, L.A (2001). Cadmium-induced changes in the growth and oxidative metabolism of pea plants. *Journal of Experimental Botany*, 52, 364, 2115-2126.

Santos de Araujo, B., Omena de Oliveira, J., Machado, A.S., Pletsch, M. (2004). Comparative studies of the peroxidases from hairy roots of *Daucus carota*, *Ipomoea batatas* and *Solanum aviculare*. *Plant Science*, 167, 1151-1157.

Sas-Nowosielska, A., Kucharski, R., Małkowski, E., Pogrzeba, M., Kuperberg, J.M., Kryński, K. (2004). Phytoextraction crop disposal-an unsolved problem. *Environmental Pollution*, 128, 373-379.

Scheckel, K.G., Hamon, R., Jassogne, L., Rivers, M., Lombi, E. (2007). Synchrotron X-ray absorption-edge computed microtomography imaging of thallium compartmentalization in *Iberis intermedia*, *Plant Soil*, 209, 51-60.

Schnoor, J.L. (1997). Phytoremediation. Technology evaluation report, Ground-Water Remediation Technologies Analysis Center, Iowa.

Schüler, G., Mithöfer, A., Baldwin, I.T., Berger, S., Ebel, J., Santos, J.G., Herrmann, G., Hölscher, D., Kramell, R., Kutchan, T.M., Maucher, H., Schneider, B., Stenzel, I., Wasternack, C., Boland, W. (2004). Coronalon: a powerful tool in plant stress physiology, *FEBS Letters.* 563, 17-22.

Schützendübel, A., Polle, A. (2002). Plant responses to abiotic stresses: heavy metal induced oxidative stress and protection by mycorrhization. *Journal of Experimental Botany,* 53, 1351-1365.

Schwartz, C., Gérard, E., Perronnet, K., Morel, J.L. (2001). Measurement of in situ phytoextraction of zinc by spontaneous metallophytes growing on a former smelter site. *The Science of the Total Environment,* 279, 215-221.

Schwartz, C., Echevarria, G., Morel, J.L. (2003). Phytoextraction of cadmium with *Thlaspi caerulescens, Plant and Soil,* 249, 27-35.

Schwitzguébel, J.-P. (2002). Hype or Hope: The Potential of Phytoremediation as an Emerging Green Technology. *Federal Facilities Environmental Journal,* 109-125.

Sekhar, K.C., Kamala, C.T., Chary, N.S., Balaram, V., Garcia, G. (2005). Potential of *Hemidesmus indicus* for phytoextraction of lead from industrially contaminated soils. *Chemosphere,* 58, 507-514.

Seth, C.S., Chaturvedi, P.K., Misra, V. (2008). The role of phytochelatins and antioxidants in tolerance to Cd accumulation in *Brassica juncea* L. *Ecotoxicology and Environmental Safety,* 71, 76-85.

Shah, K., Kumar, R.G., Verma, S., Dubey, R.S. (2001). Effect of cadmium on lipid peroxidation, superoxide anion generation and activities of antioxidant enzymes in growing rice seedlings. *Plant Science,* 161, 1135-1144.

Shah, K., Nongkynrih, J.M. (2007). Metal hyperaccumulation and bioremediation. *Biologia Plantarum,* 51, 4, 618-634.

Shanker, A.K., Cervantes, T.C., Loza-Tavera, H., Avudainayagam, S. (2005). Chromium toxicity in plants. *Environment International,* 31, 739-753.

Shao, H.-B., Chu, L.-Y., Lu, Z.-H., Kang, C.-M. (a) (2008). Primary antioxidant free radical scavenging and redox signaling pathways in higher plant cells. *International Journal of Biological Sciences,* 4, 8-14.

Shao, H.-B., Chu, L.-Y., Shao, M.-A., Cheruth, A.J., Mi, H.-M. (b) (2008). Higher plant antioxidants and redox signaling under environmental stresses. *C.R. Biologies, 331,* 433-441.

Shao, H.-B., Guo, Q.-J., Chu, L.-Y., Zhao, X.-N., Su, Z.-L., Hu, Y.-C., Cheng, J.-F. (b) (2007). Understanding of molecular mechanism of higher plant plasticity under abiotic stress. *Colloids and Surfaces B: Biointerfaces,* 54, 37-45.

Shao, H.-B., Jiang, S.-Y., Li, F.-M., Chu, L.-Y., Zhao, C.-X., Shao, M.-A., Zhao, X.-N., Li, F. (a) (2007). Some advance in plant stress physiology and their implications in the systems biology era. *Colloids and Surfaces B: Biointerfaces,* 54, 33-36.

Sharma, S. S, Dietz K.-J. (2009).The relationship between metal toxicity and cellular redox imbalance. *Trends in Plant Science,* 14, 43-50.

Sheoran, V., Sheoran, A.S., Poonia, P. (2009). Phytomining: A review. *Minerals engineering,* 22, 1007-1019.

Shri, M., Kumar, S., Chakrabarty, D., Trivedi, P.K., Mallick, S., Misra, P., Shukla, D., Mishra, S., Srivastava, S., Tripathi, R.D., Tuli, R. (2009). Effect of arsenic on growth, oxidative stress, and antioxidant system in rice seedlings. *Ecotoxicology and Environmental Safety,* 72, 4, 1102-1110.

Shtangeeva, I. (2010). Uptake of uranium and thorium by native and cultivated plants. *Journal of Environmental Radioactivity,* 101, 458-463.

Siedlecka, A., Krupa, Z. (2002). Simple method of *Arabidopsis thaliana* cultivation in liquid nutrient medium. *Acta Physiologiae Plantarum,* 24, 163-166.

Sillanpää, S., Salminen, J.-P., Lehikoinen, E., Toivonen, E., Eeva, T. (2008). Carotenoids in a food chain along a pollution gradient. *Science of the Total Environment,* 406, 247-255.

Singh, A., Sharma, R.K., Agrawal, S.B. (2008). Effects of fly ash incorporation on heavy metal accumulation, growth and yield responses of *Beta vulgaris* plants. *Bioresource Technology,* 99, 7200-7207.

Singh, N., Ma, L.Q., Srivastava, M., Rathinasabapathi, B. (2006). Metabolic adaptations to arsenic-induced oxidative stress in *Pteris vittata* L and *Pteris ensiformis* L. *Plant Science,* 170, 2, 274-282.

Sinha, S., Basant, A., Malik, A.P. Singh, K.P. (2009). Iron-induced oxidative stress in a macrophyte: A chemometric approach. *Ecotoxicology and Environmental Safety,* 72, 585-595.

Sinha, S.K., Srivastava, H.S., Tripathi, R.D. (1993). Influence of some growth regulators and cations on the inhibition of chlorophyll biosynthesis by lead in maize. *Bulletin of Environmental Contamination Toxicology,* 51, 241-246.

Škerget, M., Kotnik, P., Handolin, M., Hraš, A.R., Simonič, M., Knez, Ž. (2005). Phenols, proanthocyanidins, flavones in some plant materials and their antioxidant activities. *Food Chemistry,* 89, 191-198.

Šlesak, I., Libik, M., Karpinska, B., Karpinski, S., Miszalski, Z. (2007). The role of hydrogen peroxide in regulation of plant metabolism and cellular signalling in reponse to environmental stresses. *Acta Biochimica Polonica,* 54, 39-50.

Slováková, Ľ., Mistrík, I. (2007). Fyziologické procesy rastlín v podmienkach stresu (Physiological processes in plants under stress conditions). 1st Edition, Univerzita Komenského Bratislava.

Soudek, P., Petřík, P., Vágner, M., Tykva, R., Plojhar, V., Petrová, Š., Vaněk, T. (2007). Botanical survey and screening of plant species which accumulate 226Ra from contaminated soil of uranium waste depot. *European Journal of Soil Biology,* 43, 251-261.

Špirochová, I., Punčochářová, J., Kafka, Z., Kubal, M., Soudek, P., Vaněk, T. (2001). Studium kumulace ťežkých kovu v rostlinách. *Chemické listy,* 95, 335-336.

Srivastava, S., Mishra, S., Tripathi, R.D., Dwivedi, S., Gusta, D.K. (2006). Copper-induced oxidative stress and responses of antioxidants and phytochelatins in *Hydrilla verticillata* (L.f.) Royle. *Aquatic Toxicology,* 80, 405-415.

Stolt, J.P., Sneller, F.E.C., Bryngelsson, T., Lundborg, T., Schat, H. (2003). Phytochelatin and cadmium accumulation in wheat. *Environmental and Experimental Botany,* 49, 21-28.

Sun, Q., Wang, X.R., Ding, S.M., Yuan, X.F. (2005). Effects of exogenous organic chelators on phytochelatins production and its relationship with cadmium toxicity in wheat (*Triticum aestivum* L.) under cadmium stress. *Chemosphere,* 60, 22-31.

Sun, Q., Ye, Z.H., Wang, X.R., Wong, M.H. (2007). Cadmium hyperaccumulation leads to an increase of glutathione rather than phytochelatins in the cadmium hyperaccumulator *Sedum alfredii. Journal of Plant Physiology,* 164, 1489-1498.

Szőllősi, R., Varga, I.S., Erdei, L., Mihalik, E. (2009). Cadmium-induced oxidative stress and antioxidant mechanisms in germinating Indian mustard (*Brassica juncea* L.) seeds. *Ecotoxicology and Environmental Safety,* 72, 1337-1342.

Takahama, U. (1984). Hydrogen peroxide-dependent oxidation of flavonols by intact spinach chloroplasts, *Plant Physiology,* 74, 852-855.

Takahashi, M., Terada, Y., Nakai, I., Nakanishi, H., Yoshimura, E., Mori, S., Nishizawa, N.K. (2003). Role of Nicotianamine in the Intracellular Delivery of Metals and Plant Reproductive Development. *The Plant Cell,* 15, 1263-1280.

Tang, Y.-T., Qiu, R.-L., Zeng, X.-W., Ying, R.-R., Yu, F.-M., Zhou, X.-Y.(2009). Lead, zinc, cadmium hyperaccumulation and growth stimulation in *Arabis paniculata* Franch. *Environmental and Experimental Botany,* 66, 1, 26-134.

Tanyolaç, D., Ekmekçi, Y., Ünalan, S. (2007). Changes in photochemical and antioxidant enzyme activities in maize (*Zea mays* L.) leaves exposed to excess copper. *Chemosphere*, 67, 1, 89-98.

Taulavuori, K., Prasad, M.N.V., Taulavuori, E., Laine, K. (2005). Metal stress consequences on frost hardiness of plants at northern high latitudes: a review and hypothesis. *Environmental Pollution,* 135, 209-220.

Tewari, R.K., Kumar, P., Sharma, P.N., Bisht, S.S. (2002). Modulation of oxidative stress responsive enzymes by excess cobalt. *Plant Science,* 162, 3, 381-388.

Tordoff, G.M., Baker, A.J.M., Willis, A.J. (2000). Current approaches to the revegetation and reclamation of metalliferous wastes. *Chemosphere,* 41, 219-228.

Tsuji, N., Hirayanagi, N., Okada, M., Miyasaka, H., Hirata, K., Zenk, M. H., Miyamoto, K. (2002). Enhancement of tolerance to heavy metals and oxidative stress in *Dunaliella tertiolecta* by Zn-induced phytochelatin synthesis. *Biochemical and Biophysical Research Communications,* 293, 653-659.

Tu, C., Ma, L.Q. (2005). Effects of arsenic on concentration and distribution of nutrients in the fronds of the arsenic hyperaccumulator *Pteris vittata* L. *Environmental Pollution,* 135, 2, 333-340.

Tukendorf, A., Rauser, W. E. (1990). Changes in glutathione and phytochelatins in roots of maize seedlings exposed to cadmium. *Plant Science,* 70, 155-166.

Vamerali, T., Bandiera, M., Coletto, L., Zanetti, F., Dickinson, N.M., Mosca, G. (2009). Phytoremediation trials on metal- and arsenic-contaminated pyrite wastes (Torviscosa, Italy). *Environmental Pollution,* 157, 887-894.

Vamerali, T., Bandiera, M., Mosca, G. (2010). Field crops for phytoremediation of metal-contaminated land. A review. *Environmental Chemistry Letters,* 8, 1-17.

Vangronsveld, J., Herzig, R., Weyens, N., Boulet, J., Adriaensen, K., Ruttens, A., Thewys, T., Vassilev, A., Meers, E., Nehnevajova, E., van der Lelie, D., Mench, M. (2009). Phytoremediaiton of contaminated soils and groundwater: lessons from the field. *Environemtnal Science and Pollution Research,* 16, 765-794.

Vázquez, A., Agha, R., Granado, A., Sarro, M.J., Esteban, E., Peñalosa, J.M., Carpena, R.O. (2006). Use of white lupin plant for phytostabilization of Cd and As polluted acid soil. Water, *Air and Soil Pollution,* 177, 349-365.

Vázquez, S., Goldsbrough, P., Carpena, R.O. (2009). Comparative analysis of the contribution of phytochelatins to cadmium and arsenic tolerance in soybean and white lupin. *Plant Physiology and Biochemistry,* 47, 1, 63-67.

Ventrella, A., Catucci, L., Piletska, E., Piletsky, S., Agostiano, A. (2009). Interactions between heavy metals and photosynthetic materials studied by optical techniques. *Bioelectrochemistry,* 77, 19-25.

Vera Tomé, F., Blanco Rodríguez, P., Lozano, J.C. (2008). Elimination of natural uranium and 226Ra from contaminated waters by rhizofiltration using *Helianthus annuus* L. *Science of the total environment,* 393, 351-358.

Verma, S., Dubey, R.S. (2003). Lead toxicity induces lipid peroxidation and alters the activities of antioxidant enzymes in growing rice plants. *Plant Science,* 164, 645-655.

Vernay, P., Gauthier-Moussard, C., Hitmi, A. (2007). Interaction of bioaccumulation of heavy metal chromium with water relation, mineral nutrition and photosynthesis in developed leaves of *Lolium perenne* L. *Chemosphere,* 68, 1563-1575.

Vernay, P., Gauthier-Moussard, C., Jean, L., Bordas, F., Faure, O., Ledoigt, G., Hitmi, A. (2008). Effect of chromium species on phytochemical and physiological parameters in *Datura innoxia. Chemosphere,* 72, 5, 763-771.

Viehweger, K., Geipel, G. (2010). Uranium accumulation and tolerance in *Arabidopsis halleri* under native versus hydroponic conditions. *Environmental and Experimental Botany,* 69, 39-46.

Vögeli-Lange, R., Wagner, G. J. (1996). Relationship between cadmium, glutathione and cadmium-binding peptides (phytochelatins) in leaves of intact tobacco seedlings. *Plant Science,* 114, 11-18.

Vrettos, J.S., Stone, D.A., Brudvig, G.W. (2001). Quantifying the ion selectivity of the Ca^{2+} site in photosystem II: evidence for direct involvement of Ca^{2+} in O_2 formation. *Biochemistry,* 40, 7937-7945.

Wang, C., Zhang, S.H., Wang, P.F., Qian, J., Hou, J., Zhang, W.J., Lu, J. (2009). Excess Zn alters the nutrient uptake and induces the antioxidant responses in submerged plant *Hydrilla verticillata* (L.f.) Royle. *Chemosphere,* 76, 7, 938-945.

Warren, G.P., Alloaway, B.J., Lepp, N.W., Singh, B., Bochereau, F.J.M., Penny, C. (2003). Field trials to assess the uptake of arsenic by vegetables from contaminated soils and sol remediation with iron oxides. *The Science of the Total Environment,* 311, 19-33.

Warren, G.P., Alloway, B.J. (2003). Ecosystem Restoration, Reduction of Arsenic Uptake by Lettuce with Ferrous Sulfate Applied to Contaminated Soil. *Journal of Environmental Quality,* 32, 767-772.

Wei, S., Ma, L.Q., Saha, U., Mathews, S., Sundaram, S., Rathinasabapathi, B., Zhou, Q. (2010). Sulfate and glutathione enhanced arsenic accumulation by arsenic hyperaccumulator *Pteris vittata* L. *Environmental Pollution,* 158, 5, 1530-1535.

Whiting, S.N., Reeves, R.D., Richards, D., Johnson, M.S., Cooke, J.A., Malaisse, F., Paton, A., Smith, J.A.C., Angle, J.S., Chaney, R.L., Ginocchio, R., Jaffré, T., Johns, R., McIntyre, T., Purvis, O.W., Salt, D.E., Schat, H., Zhao, F.J., Baker, A.J.M. (2004). Research Priorities for Conservation of Metallophyte Biodiversity and their Potential for Restoration and Site Remediation. *Restoration Ecology,* 12, 1, 106-116.

Wierzbicka, M., Obidzińska, J. (1998). The effect of lead on seed imbibition and germination in different plant species. Plant Science, 137, 155-171.

Williams, L.E., Pittman, J.K., Hall, J.L. (2000). Emerging mechanisms for heavy metal transport in plants. *Biochimica et Biophysica Acta,* 1465, 104-126.

Winge, D.R., Jensen, L.T., Srinivassan, C. (1998). Metal-ion regulation of gene expression in yeast. *Current Opinion in Chemical Biology,* 2, 216-221.

Witters, N., Van Slycken, S., Ruttens, A., Adriaensen, K., Meers, E., Meiresonne, L., Tack, F.M.G., Thewys, T., Laes, E., Vangronsveld, J. (2009). Short-Rotation Coppice of Willow for Phytoremediation of a Metal-Contaminated Agricultural Area: *A Sustainability Assessment. Bioenergy Research,* 2, 144–152

Wójcik, M., Vangronsveld, J., Tukiendorf, A. (2005). Cadmium tolerance in *Thlaspi caerulescens* I. Growth parameters, metal accumulation and phytochelatin synthesis in response to cadmium. *Environmental and Experimental Botany,* 53, 151-161.

Wong, M.H. (2003). Ecological restoration of mine degraded soils, with emphasis on metal contaminated soils. *Chemosphere,* 50, 775-780.

Wood, J.M., Wang, H.-K. (1985). Microbial Resistance to Heavy Metals. In Environmental Inorganic Chemistry (Proceedings), VCH Publishers, Inc. Deerfield Beach, Florida, 487-512.

Wu, J., Hsu, F.C., Cunningham, S.D. (1999). Chelate-Assisted Pb Phytoextraction: Pb Availability, Uptake, and Translocation Constraints. *Environmental Science and Technology*, 33, 1898-1904.

Wu, F., Zhang, G., Dominy, P. (2003). Four barley genotypes respond differently to cadmium: lipid peroxidation and activities of antioxidant capacity. *Environmental and Experimental Botany*, 50, 67-78.

Wu, G., Kang, H., Zhang, X., Shao, H., Chu, L., Ruan, C. (2010). A critical review on the bio-removal of hazardous heavy metals from contaminated soils: Issues, progress, eco-environmental concerns and opportunities. *Journal of Hazardous Materials*, 174, 1-8.

Wu, L.H., Luo, Y.M., Xing, X.R., Christie, P. (2004). EDTA-enhanced phytoremediation of heavy metal contaminated soil with Indian mustard and associated potential leaching risk. *Agriculture, Ecosystems and Environment*, 102, 307-318.

Xiong, Z.-T. (1997). Bioaccumulation and physiological effects of excess lead in a roadside pioneer species *Sonchus oleraceus* L. *Environmental Pollution*, 97, 275-279.

Xiong, Z. T., Wang, H. (2005). Copper toxicity and bioaccumulation in Chinese cabbage (*Brassica pekinensis* Rupr.), *Environmental Toxicology*, 20, 188-194.

Xiong, Z.-T., Schumaker, K.S., Zhu, J.-K. (2002). Cell signaling during cold, drought and salt stress. *Plant Cell*, 165-183.

Xu, J., Yang, F., Chen, L., Hu, Y., Hu, Q. (2003). Effect of Selenium on Increasing the Antioxidant Activity of Tea Leaves Harvested during the Early Spring Tea Producing Season. *Journal of Agricultural and Food Chemistry*, 51, 1081-1084.

Xu, W., Li, W., He, J., Singh, B., Xiong, Z. (2009). Effects of insoluble Zn, Cd, and EDTA on the growth, activities of antioxidant enzymes and uptake of Zn and Cd in *Vetiveria zizanioides*. *Journal of Environmental Sciences*, 21, 2, 186-192.

Yadav, S. K. (2010). Heavy metals toxicity in plants: An overview on the role of glutathione and phytochelatins in heavy metal stress tolerance of plants. *South African Journal of Botany*, 76, 2, 167-179.

Yang, X., Feng, Y., He, Z., Stoffella, P.J. (2005). Molecular mechanisms of heavy metal hyperaccumulation and phytoremediation. *Journal of Trace Elements in Medicine and Biology*, 18, 339-353.

Yanqun, Z., Yuan, L., Jianjun, C., Haiyan, C., Li, Q., Schvartz, C. (2005). Hyperaccumulation of Pb, Zn and Cd in herbaceous grown on lead–zinc mining area in Yunnan, China. *Environment International*, 31, 755-762.

Yoon, J., Cao, X., Zhou, O. (2006). Accumulation of Pb, Cu and Zn in native plants growing on a contaminated Florida site. *The Science of the Total Environment,* 368, 456-464.

Yordanov, I., Velikova, V., Tsonev, T. (2000). Plant responses to drought, acclimation, and stress tolerance. *Photosynthetica* 38, 171-186.

Zabłudovska, E., Kowalska, J., Jedynak, Ł. Wojas, S., Skłodowska, A., Antosiewicz, D.M. (2009). Search for a plant for phytoremediation – What can we learn from field and hydroponic studies? *Chemosphere,* 77, 301-307.

Zaman, M.S., Zereen, F. (1998). Growth Responses of Radish Plants to Soil Cadmium and Lead Contamination. *Bulletin of Environmental Contamination Toxicology,* 61, 44-50.

Zeid, I.M. (2001). Responses of *Phaseolus vulgaris* to chromium and cobalt treatments, *Biologia Plantarum,* 44, 111-115.

Zenk, M.H. (1996). Heavy metal detoxification in higher plants – a review. *Gene* 179, 21-30.

Zhang, F.-Q., Wang, Y.-S., Lou, Z.-P., Dong, J.-D. (2007). Effect of heavy metal stress on antioxidant enzymes and lipid peroxidation in leaves and roots of two mangrove plant seedlings (*Kandelia candel* and *Bruguiera gymnorrhiza*). *Chemosphere,* 67, 44-50.

Zhang, H., Jiang, Y., He, Z., Ma, M. (2005). Cadmium accumulation and oxidative burst in garlic (*Allium sativum*). *Journal of Plant Physiology,* 162, 977-984.

Zhang, H., Xu, W., Guo, J., He, Z., Ma, M. (2005). Coordinated responses of phytochelatins and metallothioneins to heavy metals in garlic seedlings. *Plant Science,* 169, 1059-1065.

Zhang, J., Jia, W., Yang, J., Ismail, A.M. (2006). Role of ABA in integrating plant response to drought and salt stresses. *Field Crops Research,* 97, 111-119.

Zhang, J., Kirkham, M.B. (1996). Enzymatic responses of the ascorbate-glutathione cycle to drought in *Sorghum* and sunflower plants. *Plant Science.* 113, 139-147.

Zhang, W., Cai, Y., Downum, K.R., Ma, L.Q. (2004). Thiol synthesis and arsenic hyperaccumulation in *Pteris vittata* (Chinese brake fern). *Environmental Pollution,* 131, 337-345.

Zhang, Z., Gao, X., Qiu, B. (2008). Detection of phytochelatins in the hyperaccumulator *Sedum alfredii* exposed to cadmium and lead. *Phytochemistry,* 69, 4, 911-918.

Zhao, F.J., Hamon, R.E., McLaughlin, M.J. (2001). Root exudates of the hyperaccumulator *Thlaspi caerulescens* do not enhance metal mobilization. *New Physiologist,* 151, 613-620.

Zhou, M., Ze, Y., Li, N., Duan, Y., Chen, T., Lin, C., Hong, F. (2009). Cerium relieving the inhibition of photosynthesis and growth of the spinach caused by lead. *Journal of Rare Earths,* 27, 5, 864-869.

Zhou, W., Qiu, B. (2007). Effects of cadmium hyperaccumulation on physiological characteristics of *Sedum alfredii* Hance (Crassulaceae). *Plant Science,* 169, 737-745.

Zhu, Z., Wei, G., Li, J., Qian, Q., Yu, J. (2004). Silicon alleviates salt stress and increases antioxidant enzymes activity in leaves of salt-stressed cucumber (*Cucumis sativus* L.). *Plant Science,* 167, 527-533.

Zurita, F., De Anda, J., Belmont, M.A. (2009). Treatment of domestic wastewater and production of commercial flowers in vertical and horizontal subsurface-flow constructed wetlands. *Ecological Engineering,* 35, 861-869.

INDEX